Fundamentals of Non-Thermal Processes for Food Preservation

The textbook comprises ten chapters, which are written in a simple but scientific language, encompassing all the non-thermal treatments in-depth, from basic concepts to technological advances. This book provides complete study material in a single source including such pedagogical features as multiple choice questions, solved numerical problems, and short questions. The book begins with a general introduction to the evolution of the non-thermal technique for food preservation. The fundamental mechanism of non-thermal inactivation of microorganisms and enzymes is discussed. In the following chapters, eight non-thermal techniques have been discussed in detail.

Fundamentals of Non-Thermal Processes for Food Preservation

Snehasis Chakraborty and Rishab Dhar

CRC Press
Taylor & Francis Group
Boca Raton London New York

CRC Press is an imprint of the
Taylor & Francis Group, an **informa** business

First edition published 2023
by CRC Press
6000 Broken Sound Parkway NW, Suite 300, Boca Raton, FL 33487-2742

and by CRC Press
4 Park Square, Milton Park, Abingdon, Oxon, OX14 4RN

CRC Press is an imprint of Taylor & Francis Group, LLC

Library of Congress Cataloging-in-Publication Data

Names: Chakraborty, Snehasis, author. | Dhar, Rishab, author.
Title: Fundamentals of non-thermal processes for food preservation /
Snehasis Chakraborty and Rishab Dhar.
Description: First edition. | Boca Raton : CRC Press, 2023. | Includes
bibliographical references.
Identifiers: LCCN 2022010750 | ISBN 9781032059099 (hardback) |
ISBN 9781032040592 (paperback) | ISBN 9781003199809 (ebook)
Subjects: LCSH: Food--Preservation.
Classification: LCC TX601 .C45 2023 | DDC 641.4--dc23/eng/20220630
LC record available at https://lccn.loc.gov/2022010750

ISBN: 9781032059099 (hbk)
ISBN: 9781032040592 (pbk)
ISBN: 9781003199809 (ebk)

DOI: 10.1201/9781003199809

Typeset in Adobe Caslon Pro
by KnowledgeWorks Global Ltd.

Contents

PREFACE   xi

ABOUT THE AUTHORS   xiii

CHAPTER 1   NON-THERMAL PROCESSES IN FOOD
PRESERVATION   1

1.1	Principles of Food Preservation	1
1.2	Inception of Non-Thermal Processing	2
1.3	Various Non-Thermal Techniques	4
1.4	Principal Actions of Non-Thermal Processing	6
1.5	Status and Trends of Non-Thermal Technologies	7
	1.5.1 High-Pressure Processing	8
	1.5.2 Pulsed Electric Field	8
	1.5.3 Pulsed Light and Ultraviolet Technology	9
	1.5.4 Power Ultrasound Processing	9
	1.5.5 Cold Plasma Technology	10
	1.5.6 Ozone Processing	10
	1.5.7 Irradiation	10
	1.5.8 Oscillating Magnetic Field	11
1.6	Summary	12
1.7	Multiple Choice Questions	13
1.8	Short Answer Type Questions	14
1.9	Descriptive Questions	15
	Suggested Readings	15
	Answers for MCQs (sec. 1.7)	15

CHAPTER 2   **HIGH-PRESSURE PROCESSING**   17
2.1   Principle   17
2.2   Operation   17
    2.2.1   Batch Mode   18
    2.2.2   Semicontinuous Mode   19
2.3   Components of High-Pressure Processing System   20
    2.3.1   Pressure Vessel   20
    2.3.2   Pressure Generation System   21
    2.3.3   Pressure-Transmitting Medium   21
    2.3.4   Temperature Controller   22
    2.3.5   Heat of Compression   22
    2.3.6   Processing Time   23
    2.3.7   Suitable Packaging Materials   24
2.4   Mechanism of Quality Changes in Food   24
    2.4.1   Mechanism Microbial Inactivation   24
    2.4.2   Mechanism of Inactivating Spoilage Enzymes   25
    2.4.3   Mechanism of Nutritional Quality Change   26
2.5   Critical Parameters for Process Design   26
    2.5.1   Type of Product   26
    2.5.2   Target Microorganism or Enzyme   26
    2.5.3   Equipment and Mode of Operation   27
2.6   Practical Example   29
2.7   Potential and Challenges   30
2.8   Summary   30
2.9   Solved Numerical   31
2.10   Multiple Choice Questions   37
2.11   Short Answer Type Questions   38
2.12   Descriptive Questions   39
2.13   Numerical Problems   39
References   40
Suggested Readings   40
Answers for MCQs (sec. 2.10)   41

CHAPTER 3   **PULSED ELECTRIC FIELD PROCESSING**   43
3.1   Principles and Operation   43
3.2   Mechanism of Quality Changes in Food   44
    3.2.1   Microbial Inactivation   44
    3.2.2   Inactivation of Spoilage Enzymes   47
    3.2.3   Effect of Pulsed Electric Field on Nutritional Quality   47
3.3   Equipment and Critical Processing Factors   48
    3.3.1   Type of Product   49
    3.3.2   Target Microorganism   49
    3.3.3   Equipment and Mode of Operation   50
3.4   Practical Example   52
3.5   Challenges   53
3.6   Summary   54

3.7	Solved Numerical	55
3.8	Multiple Choice Questions	59
3.9	Short Answer Type Questions	61
3.10	Descriptive Questions	61
3.11	Numerical Problems	62
	References	62
	Suggested Readings	63
	Answers for MCQs (sec. 3.8)	63

CHAPTER 4   ULTRAVIOLET AND PULSED LIGHT TREATMENT   65

4.1	Principle and Operation	65
4.2	Mechanism of Quality Changes in Food	66
	4.2.1 Microbial Inactivation	66
	4.2.2 Effect on Spoilage Enzyme and Nutritional Quality	68
4.3	Equipment Functioning	68
4.4	Critical Processing Factors	70
	4.4.1 Type of Product	70
	4.4.2 Relative Positioning of the Product	71
	4.4.3 Target Microorganism or Enzyme	71
	4.4.4 Equipment and Mode of Operation	72
4.5	Practical Example	74
4.6	Challenges	75
4.7	Summary	76
4.8	Solved Numerical	77
4.9	Multiple Choice Questions	79
4.10	Short Answer Type Questions	82
4.11	Descriptive Questions	82
4.12	Numerical Problems	83
	References	83
	Suggested Readings	83
	Answers for MCQs (sec. 4.9)	84

CHAPTER 5   POWER ULTRASOUND PROCESSING   85

5.1	Principles	85
5.2	Mechanism of Quality Changes in Food	87
	5.2.1 Microbial Inactivation	87
	5.2.2 Inactivation of Spoilage Enzymes	88
	5.2.3 Effect on Nutritional Quality	89
5.3	Equipment and Its Operation	89
5.4	Critical Processing Factors	92
	5.4.1 Type of Medium	92
	5.4.2 Target Microorganism and Enzyme	92
	5.4.3 Equipment and Mode of Operation	93
5.5	Practical Example	94
5.6	Challenges	95
5.7	Summary	95
5.8	Solved Numerical	96

5.9	Multiple Choice Questions	100
5.10	Short Answer Type Questions	103
5.11	Descriptive Questions	103
5.12	Numerical Problems	103
	References	104
	Suggested Readings	104
	Answers for MCQs (sec. 5.9)	104
CHAPTER 6	**COLD PLASMA PROCESSING**	**105**
6.1	Principles	105
6.2	Mechanisms of Changes in Quality Attributes of Food	107
6.3	Equipment and Its Operation	109
6.4	Critical Processing Parameters	112
	6.4.1 Type of Product	112
	6.4.2 Target Microorganism or Enzyme	113
	6.4.3 Equipment and Mode of Operation	113
6.5	Practical Example	114
6.6	Challenges	114
6.7	Summary	115
6.8	Solved Numerical	116
6.9	Multiple Choice Questions	119
6.10	Short Answer Type Questions	121
6.11	Descriptive Questions	121
6.12	Numerical Problems	122
	References	123
	Suggested Readings	123
	Answers for MCQs (sec. 6.9)	123
CHAPTER 7	**OZONE PROCESSING**	**125**
7.1	Principles	125
7.2	Mechanisms of Quality Changes in Food	127
7.3	Equipment and Its Operation	128
7.4	Critical Processing Factors	130
	7.4.1 Type of Product	130
	7.4.2 Target Microorganism or Enzyme	130
	7.4.3 Equipment and Mode of Operation	131
7.5	Practical Example	131
7.6	Challenges	132
7.7	Summary	133
7.8	Solved Numerical	134
7.9	Multiple Choice Questions	138
7.10	Short Answer Type Questions	140
7.11	Descriptive Questions	140
7.12	Numerical Problems	140
	References	141
	Suggested Readings	142
	Answers for MCQs (sec. 7.9)	142

Chapter 8	**IRRADIATION**	143
	8.1 Principle	143
	8.2 Mechanism of Lethality	146
	8.2.1 Microbial Inactivation	146
	8.2.2 Spoilage Enzymes Inactivation	147
	8.3 Effect of Irradiation on Nutritional Quality	148
	8.4 Critical Processing Factors	149
	8.4.1 Type of Product	149
	8.4.2 Target Microorganism or Enzyme	150
	8.4.3 Irradiation Equipment and Process Parameters	151
	8.5 Dose of Irradiation	155
	8.6 Practical Example	155
	8.7 Challenges	156
	8.8 Summary	157
	8.9 Solved Numerical	158
	8.10 Multiple Choice Questions	161
	8.11 Short Answer Type Questions	163
	8.12 Descriptive Questions	163
	8.13 Numerical Problems	164
	References	164
	Suggested Readings	164
	Answers for MCQs (sec. 8.10)	165
Chapter 9	**OSCILLATING MAGNETIC FIELD PROCESSING**	167
	9.1 Principle and Operation	167
	9.2 Mechanism of Lethality	169
	9.2.1 Microbial Inactivation	169
	9.2.2 Spoilage Enzymes Inactivation	172
	9.3 Effect of OMF on Nutritional Quality and Freezing of Food Products	173
	9.4 Equipment	174
	9.5 Critical Processing Factors	176
	9.5.1 Type of Product	176
	9.5.2 Target Microorganism or Enzyme	176
	9.5.3 Equipment and Mode of Operation	177
	9.6 Practical Example	177
	9.7 Challenges	179
	9.8 Summary	180
	9.9 Solved Numerical	181
	9.10 Multiple Choice Questions	184
	9.11 Short Answer Type Questions	186
	9.12 Descriptive Questions	186
	9.13 Numerical Problems	186
	References	187
	Suggested Readings	187
	Answers for MCQs (sec. 9.10)	188

CHAPTER 10 COMMERCIAL ASPECTS AND CHALLENGES 189
    10.1 Potential and Limitations 189
    10.2 Sustainability 189
        10.2.1 Social Impacts 194
        10.2.2 Environmental Impacts 195
        10.2.3 Economic Aspects 196
    10.3 Commercialization Aspects 197
        10.3.1 Consumer Response 197
        10.3.2 Market Status 198
    10.4 Packaging Challenges 199
    10.5 Regulatory and Legislative Issues 200
    10.6 Challenges and Research Needs 203
        10.6.1 Equipment Cost 203
        10.6.2 Maintenance 204
        10.6.3 Scale-up Trials 204
        10.6.4 Market and Consumer Acceptance 204
        10.6.5 In-depth Understanding of Lethality 205
        10.6.6 Sustainability Aspect 205
        10.6.7 Type of Product and Matrix Properties 205
        10.6.8 Shelf Life and Stability 205
    10.7 Strategy for Process Optimization 206
    10.8 Summary 207
    10.9 Multiple Choice Questions 208
    10.10 Short Answer Type Questions 210
    10.11 Descriptive Questions 210
    References 211
    Suggested Readings 211
    Answers for MCQs (sec. 10.9) 212

INDEX 213

Preface

Fundamentals of Non-Thermal Processes of Food Preservation covers a significant portion of novel or unconventional food processing and preservation technologies, mainly in the form of an elective for both Bachelor's and Master's students in Food Science, Food Technology, or Food Engineering. This textbook explores the fundamental aspects of various non-thermal techniques for food preservation to facilitate both classroom teaching as well as learning processes. This book has ten chapters, written in a scientific but straightforward language, encompassing all the non-thermal treatments in depth, from basic concepts to technological advances.

This textbook begins with a general introduction to the evolution of the non-thermal techniques for food preservation. In the following chapters, a set of eight non-thermal techniques have been discussed in detail. The techniques are high-pressure processing, pulsed electric field processing, ultraviolet and pulsed light treatment, ultrasound treatment, cold plasma treatments, ozonization, gamma irradiation, and oscillating magnetic field processing. For these eight chapters, the general structure is principle and operation, mechanism of microbial inactivation, enzyme, and nutritional quality, equipment and process parameters, typical industrial examples, issues, and challenges. It is followed by a bullet-point summary of the entire technique. There are numerical problems with solutions, multiple choice questions, short and long answer questions for each chapter. Finally, a few unsolved numericals are provided as home assignments.

We believe this book can be a ready reckoner for undergraduate and postgraduate students in Food Science and Technology or Food Engineering to learn various non-thermal techniques for food preservation. We shall be happy to receive suggestions for improving the first edition of this textbook to incorporate into the next edition.

Snehasis Chakraborty & Rishab Dhar
Mumbai, India

About the Authors

Dr. Snehasis Chakraborty is an Assistant Professor at the Institute of Chemical Technology, Mumbai, India. His research area includes non-thermal and advanced thermal processing of foods, process optimization, kinetic modeling, shelf-life study, synbiotic foods, and sensory analysis. He is the editorial board member of the *Journal of Food Engineering and Technology*, Tech Reviews.

Mr. Rishab Dhar is a doctoral candidate from the Food Engineering and Technology Department, Institute of Chemical Technology, Mumbai, India. His research focuses on kinetic modeling under thermal and non-thermal treatment, process optimization, fuzzy optimization of the sensory analysis, and the enzyme-assisted juice extraction process.

1

NON-THERMAL PROCESSES IN FOOD PRESERVATION

1.1 Principles of Food Preservation

Food preservation can be broadly described as employing any processing on food products to extend their shelf life. The shelf life of the food refers to the duration from its harvesting or processing till it is considered spoiled. Spoiled food is primarily characterized as the microbial population in that product crossing the threshold limit set by any regulatory body. For instance, a fruit product is considered spoiled when the microbial population (mesophilic bacteria count) crosses 10^6 colony-forming units (CFU)/g of the sample (**Fig. 1.1**). Before reaching the threshold limit adjudged as spoiled food, food processing is employed to increase the period. Food processing can be described as applying continuous or one-off stress on the microorganism to hamper the two primary functions, viz. survival and multiplication. Out of the two primary functions, survival is the priority for a microorganism due to mortality. This stress can be applied in the form of energy like heat energy. In the non-thermal treatments, stress is applied to the microorganism in energy other than heat.

In most cases, unprocessed food contains many microorganisms. To increase the shelf life of the food, the intended processing can follow any of three principles, viz. inactivation, inhibition, and restriction (**Fig. 1.1**). Typically, inactivation means killing the microorganism, whereas inhibition refers to suppressing their growth. Inactivation targets microbial mortality, whereas inhibition aims to stop or control the secondary action – multiplication. The restriction principle works on forbidding the entry of microorganisms into the food system, such as packaging. Any kind of processing follows either or combination of these three principles such as pasteurized and packaged milk accommodates all three together. Heat treatment of milk at 72°C for 15 s inactivates the pathogens. The post-processing growth

DOI: 10.1201/9781003199809-1

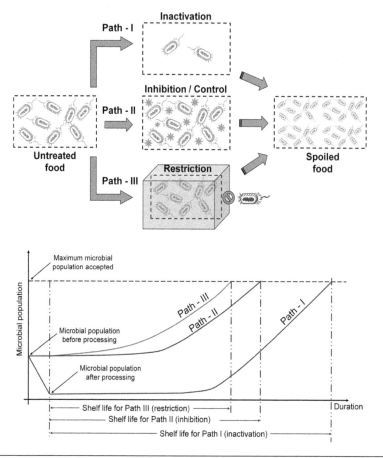

Figure 1.1   Three fundamental principles of food preservation to extend the shelf life of food.

of spoilage microorganisms is controlled or inhibited by chilled or low-temperature storage, whereas the entry of further contamination is prevented by outside package.

1.2 Inception of Non-Thermal Processing

Conventionally, thermal or heat treatment provides an extended shelf life for food processing. Examples include thermal pasteurization, sterilization through retort processing, canning, and so on. The thermal treatment combines the time and temperature when the food is processed. For instance, the batch thermal pasteurization of 95°C/5 min means that the food product is kept at 95°C for 5 min to prevent microbial spoilage. The magnitude of microbial inactivation increases with the treatment time and temperature. However, an intense heat treatment leads to a loss in nutrients besides ensuring microbial safety. In this sense, if we keep increasing the intensity (time-temperature combination) of the heat treatment, the product will be safer to consume or have an extended shelf life.

On the other hand, there will be an increase in nutrient (like vitamin) loss. However, consumers want to have a nutritionally superior product. In this sense, there is a need for an alternative processing technique in which food will be nutritionally rich while satisfying the microbial safety aspect. Nutrient degradation in thermal processing depends on both temperature and time. An increase in either time or temperature can lead to a higher loss in the desired nutrient. As the conventional technique combines time and temperature, two options are there to improve further. The treatment time can be lessened by increasing the magnitude of temperature or accelerating the heat transfer rate, such as volumetric heating methods.

Minimizing undesirable changes can be achieved through more effective process control, high-temperature short time (HTST) techniques, such as aseptic processing, or newer technologies. For example, in the case of infrared and dielectric heating, such as microwaves and ohmic heating, volumetric heating leads to a minimal treatment time. These techniques are called advanced thermal processing (**Fig. 1.2**).

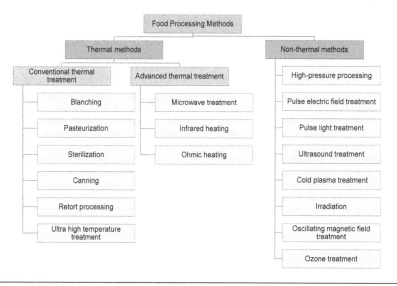

Figure 1.2 Categorization of various food processing techniques.

The other way to find an alternative food processing technique is to replace heat energy with any other form of energy, resulting in a similar microbial inactivation. This is called the non-thermal processing of food. In the non-thermal treatments, the food is processed without heat energy while ensuring microbial inactivation and nutrient retention in the product.

The primary objectives of non-thermal food processing are as follows:

a. To ensure the microbial safety of the processed food
b. To inactivate the undesirable enzymes in the food
c. To provide an extended shelf life of processed food products
d. To retain the natural color, flavor, and sensory profile of the processed food
e. To provide the food similar to the untreated product by retaining the thermosensitive compounds in the processed food product

The examples of non-thermal processing include high-pressure processing (HPP), pulsed electric field (PEF), pulsed light (PL), ultrasound (US), cold plasma (CP) treatments, irradiation, oscillating magnetic field (OMF), and ozone treatment (**Fig. 1.2**).

1.3 Various Non-Thermal Techniques

Pasteurization, sterilization, and other food processing operations are traditionally followed using thermal techniques. In these techniques, heat is transferred to food materials through conduction and convection; such ways of processing have problems of heat loss before utilization, scaling in heat concentrated areas (further reducing the heat transfer efficiency), and overheating at the food contact surface. On the contrary, in non-thermal food processing, the heat energy is replaced by any other energy such as pressure, electric field, light, sound, magnetism, etc. A summary of the principles of these techniques has been presented in **Table 1.1**.

In the case of HPP, the packaged food is placed inside a liquid medium, and the hydrostatic pressure is generated isostatically across the food and the medium. An electric field is employed in pulses on the food in PEF treatment. The food is placed in between two high-voltage electrodes. For PL treatment, the food is exposed to a high-intensity light released in highly concentrated bursts of more powerful energy.

Table 1.1 Summary of the Principles of Various Non-Thermal Food Preservation Methods

METHODS	DESCRIPTION	PRINCIPLE ACTION ON MICROORGANISM	APPLICATION
High-pressure processing	Food is compressed up to 600 MPa isostatically inside a pressure transmitting medium for a few minutes	• Permeabilization of cell membrane • Denaturation of cell wall enzymes	Solid and liquid (packaged) foods
Pulsed electric field	Food is placed between two electrodes with high-electric voltage discharge pulses (≤ 80 kV/cm) for fractions of seconds	• Electrical breakdown and electroporation of cell membrane	Liquid (mostly) and solid foods
Pulsed light treatment	Food is exposed to light flashes (1–20 per second) of high energy (10 $J \cdot cm^{-2}$) for a few seconds	• UV induced mutations in DNA and RNA • Thermal damage due to localized heating	Solid (mostly) and non-turbid liquids
Ultrasound treatment	Food is placed inside a power ultrasound condition (18–100 kHz; ≥ 1 $W \cdot cm^{-2}$) for a few minutes	• Localized heating and production of free radicals due to acoustic cavitation	Liquid (mostly) and solid foods
Cold plasma treatment	Food is exposed to the dielectric discharge of gaseous atoms placed between two electrodes	• Oxidative damage on the microbial cells by reactive species	Solid foods
Irradiation	The food is exposed to ionizing radiations for a few seconds	• Mutations at the genetic level • Disruption of the cytoplasmic membrane	Solid food (packaged)
Oscillating magnetic field	The food is exposed within an OMF with 5–10 T magnetic field (5–500 kHz) for 25–100 ms	• Denaturation of membrane enzymes and metabolic proteins • Breakdown of covalent bonds in DNA	Solid and liquid (packaged) foods
Ozonation	Ozone is injected into the medium in contact with food	• Oxidation of sulfhydryl groups and amino acids and polyunsaturated fatty acids • Penetration of acid peroxides into the genetic material	Solid and liquid foods

US treatment employs a more potent form of sound (> 5 W/cm^2) at a lower frequency (around 40 kHz), which has the most significant impact on microbes present in food. The irradiation process involves exposing the food, either prepackaged or in bulk, to a predetermined level of ionization radiation. In the case of OMF, the prepackaged food is placed between the magnetic coils or within the coil, and it is subjected to 5–10 T magnetic field having 5–500 kHz frequency. CP treatment involves the dielectric discharge of gaseous atoms placed between two electrodes under high-voltage differences. These high-energy atoms are exposed to the food surface, causing damage to the microbial cells. In the case of ozonization, the food sample is either directly exposed to the ozone gas or indirectly through ozonated water.

Although an inevitable temperature increase is possible during some of these techniques, the possibility of temperature rise is minimal for all these processing techniques. For example, the heat of compression of the medium resulted in 2–3°C rise per 100 MPa increase during HPP. High-intensity PL may lead to an increase in the surface temperature of the food. Similarly, the mechanical energy due to shear is partly converted to heat energy during US treatment. However, in all the cases, the temperature rise remains below 20°C when industrial operations are concerned. In turn, the thermal lethality remains minimal than the non-thermal counterpart.

1.4 Principal Actions of Non-Thermal Processing

Most non-thermal techniques work on the inactivation of microorganisms in the food. These techniques try to disturb the homeostasis of microorganisms. Homeostasis tends to maintain a balance inside the microbial environment to provide stability. When a microorganism is put under any kind of stress or unfavorable environment, such as heat, low pH, high sugar and acid concentration, pressure, light, magnetic field, electric field, etc., the homeostasis responds accordingly to minimize the influence of that stress on survivability. Application of thermal stress targets multiple sites inside the cell. These include outer and cytoplasmic membrane, peptidoglycan wall, RNA, ribosomes, and enzymes, such as RNA polymerase. In this way, microorganisms lose their ability to survive. However, microbial inactivation under non-thermal stress varies with the counterpart by thermal stress. The principal action of microbial inactivation of various non-thermal treatments has been summarized in **Table 1.1**.

HPP affects the microbial cell at multiple sites. Change in membrane fluidity, denaturation of membrane enzymes and RNA isomerase, and shift in cytoplasmic pH are the significant phenomena that occur inside the cell during HPP, which lead to microbial inactivation. In the case of PEF, due to electrical breakdown, electroporation of cell membrane happens, and survival of the cell becomes difficult. PL exposure causes mutations in DNA and RNA structures induced by the ultraviolet light in the spectrum. This is called the photochemical effect. Besides, the photothermal effect from the infrared portion of the spectrum can induce localized heating in the cell. The principal action of the US is acoustic cavitation, leading to localized heating and the production of free radicals responsible for microbial inactivation. The reactive species in cold plasma inflict direct oxidative damage on the outer surface of microbial cells. Irradiation induces mutations at the genetic level and disruption of the cytoplasmic membrane. When the food is exposed to the OMF, the bond between ions and proteins loosens, leading to the denaturation of enzymes and other metabolic proteins. Besides, this leads to the breakdown of covalent bonds in DNA inside the cell. Mechanism of disinfection using ozone gas includes the oxidation of sulfhydryl groups, amino acids of enzymes, and proteins. Oxidation of polyunsaturated fatty acids to acid peroxides leads to cell lysis. The damage may further be incurred to genetic material once the cell wall and cell membrane are penetrated.

The impact of non-thermal stress on the nutritional compound and phytochemicals is encouraging. Phytochemicals are believed to have health benefits and antioxidant activities. Typically, these phytochemicals degrade at a higher temperature. The non-thermal stress usually influences the weaker interactions like van der Waals interactions, hydrogen bonds, etc., and it minimally affects covalent bonds. In turn, the bioactive compounds and phytochemicals remain unaffected, leading to a fresh product.

1.5 Status and Trends of Non-Thermal Technologies

Consumers are more aware of and demand minimally processed and nutritionally rich stable foods in recent decades. The market for non-thermal food processing was valued at USD 760.7 million in 2016.

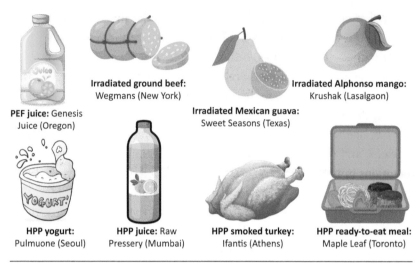

PEF juice: Genesis Juice (Oregon)

Irradiated ground beef: Wegmans (New York)

Irradiated Mexican guava: Sweet Seasons (Texas)

Irradiated Alphonso mango: Krushak (Lasalgaon)

HPP yogurt: Pulmuone (Seoul)

HPP juice: Raw Pressery (Mumbai)

HPP smoked turkey: Ifantis (Athens)

HPP ready-to-eat meal: Maple Leaf (Toronto)

Figure 1.3   Commercial non-thermal (HPP, Irradiation, and PEF) processed food products.

It is projected to reach USD 123 billion by 2023, at a compound annual growth rate (CAGR) of 8.4% from 2017. Industrially, non-thermal processing is applied for various food products, including fruit and vegetable products, juice and beverages, dairy products, meat and seafood products, cereal-based formulations, ready-to-eat (RTE) meals, etc. A set of high-pressure processed, PEF treated, and irradiated products available on the market worldwide have been shown in **Fig. 1.3**.

1.5.1 High-Pressure Processing

In HPP, food is subjected to pressure ranging from 50 to 1000 MPa. HPP is also well-known as hyperbaric pressure processing or ultra-HPP, or pascalization. Adrien Certes reported the ability of HPP to kill microorganisms in water in 1883. In 1899, Bert Hite first applied this technique to food products, and spoilage microorganisms were inactivated. Later in the 1980s, the food industries realized their commercial advantages. In the late 1980s in Japan, acid food products like yogurt and strawberry jam were first processed through HPP, and its pasteurization ability was established.

1.5.2 Pulsed Electric Field

In 1967, Sala & Hamilton were the first to execute experiments to understand the effect of the PEF on bacteria and yeasts. They found that microbial inactivation is not happening because of the products formed during electrolysis. Later in the 1980s, the membrane rupture theory came into the picture, which happens to microbial cells after strong PEF. It was observed that PEF treatment could improve the permeability of cells and thus be capable of microbial inactivation in food material (mainly liquid and semisolid) without significant effects on other food qualities. The application of PEF to improve the shelf life of apple cider products was one of the first commercial practices in the USA (2005). Apart from microbial and spoilage enzyme inactivation, PEF has also been utilized for cell hybridization (in biotechnology and genetics), achieving specific product transformations, facilitating better cellular contents, extraction of bioactive compounds, and wastewater treatment.

1.5.3 Pulsed Light and Ultraviolet Technology

Pulsed light (PL) and ultraviolet (UV) are electromagnetic radiation-based technologies that come under the umbrella of non-thermal processing techniques for food products. UV-based decontamination has been utilized since the 1900s, majorly for wastewater and drinking water treatment. PL, also known as pulsed UV, is a polychromatic light with higher penetration and four to six times more efficient in microbial inactivation than continuous UV but requires a shorter exposure time to avoid product heating. PL is comparatively a newer technique that came into the picture since the late 1970s, which is capable of disinfection and preserving the quality attributes of the food product better than conventional thermal pasteurization. UV and PL have also been used to treat food surfaces and contact surfaces.

1.5.4 Power Ultrasound Processing

US waves are sound waves that require a medium to travel. US possesses frequencies above 16 kHz, which is not audible to human ears. The first research work in US technology is entitled 'The chemical effects of high-frequency sound waves'. It was a preliminary survey published in 1927. The use of power US became established in the processing industries by the 1960s. Numerous US applications in food processing include ultrasonic cutting, defoaming, emulsification, degassing, extraction, filtration, sterilization, pasteurization, meat tenderization, thawing, etc.

1.5.5 Cold Plasma Technology

Plasma is considered the fourth state of matter, which can be viewed as a gas of ions, electrons, free radicals, atoms, and molecules at the ground or excited state. It also exists in nature in the form of lightning and polar lights. In 1857, Ernst Siemen studied electrical discharge using dielectric barrier discharge (DBD), one of the earliest studies associated with plasma technology. Later in 1928, Irving Langmuir compared such charged particles in a neutral gas-like fluid medium. CP has numerous food-based applications, such as microbial decontamination, food contact surface disinfection, food package disinfection, modifications of food components, hydrogenation of edible oil, degradation of pesticides, and food allergens.

1.5.6 Ozone Processing

Trioxygen (O_3) is an inorganic molecule known as ozone. It is a gas, either colorless or pale blue, and has a pungent smell, thus derived its name from the Greek word *ozein* (means 'to smell'). Ozone is less stable than diatomic oxygen (O_2) and possesses high oxidizing power. Ozone was first discovered by a German-Swiss chemist, Christian Schönbein, in 1840. Since 1906, the Ozonation process has been utilized to purify water in Europe. The ozone gas has high reactivity, good penetrability, and spontaneous decomposition into numerous free radicals and nontoxic oxygen and leaves no residue. All these properties make it an excellent alternative for microbial decontamination of food products (like water, meat, poultry, fish, spices, fruits, vegetables, beverages, dairy, etc.), surface disinfection, deodorization, degradation of pesticides or harmful compounds, and a much better replacement to the commonly used chemical-based disinfection.

1.5.7 Irradiation

Irradiation is the emission and transmission of energy through food products. This is a physical method of food preservation that can be applied to packaged food. Ionizing radiations (wavelength ≤ 2000 A) such as electron beam emissions, gamma rays, and X-rays are primarily used for food preservations. The gamma rays can be produced from the radiation of Cobalt 60 and Cesium 137 (these two are the approved ones). The research in irradiation started with the discovery of X-rays by Wilhelm Conrad Röntgen in 1895. In 1904, the bactericidal effects of irradiation were studied by Samuel Prescott. The applications of irradiation in food preservation were patented in 1906 (United Kingdom) and 1918 (the United States of America). Further, in 1980, the Joint Expert Committee by the World Health Organization (WHO), Food and Agriculture Organization (FAO) of the United Nations, and International Atomic Energy Agency (IAEA) cleared the use of irradiation for food preservation with a maximum dosage of 10 kGray. However, the use of radioactive sources for food preservation is still being a concern in some countries.

1.5.8 Oscillating Magnetic Field

Food preservation by OMFs is relatively new. This technique keeps the food within OMFs generated with alternating current electromagnets. The intensity varies periodically according to the frequency and type of wave in the magnet. This eventually inactivates the microorganisms. The concept of magnetism and magnetic field was quantified in 1865 by James Clerk Maxwell. Galileo Ferraris introduced the idea of rotating magnetic fields in 1885. Nicholas Tesla had independently applied the principle of rotating magnetic fields in 1887. However, microbial inhibition by magnetic field was first observed by Vincent F. Gerencser in 1962. Later in 1984, PurePulse Technologies Inc. (Maxwell Technologies Inc.) patented the concept of deactivation of microorganisms by an OMF. The use of OMF in food preservation is still in the research phase.

Each of these eight non-thermal techniques has been detailed further in the following chapters. These technologies promise to maintain the critical balance between the safety and marketability of a new generation of foods. Some of these technologies have been optimized for the cold pasteurization of foods, and such products are now commercially available.

However, using a non-thermal preservation method for a particular food depends upon various criteria like the equipment cost, scale, and cost of production, type of food product, intended shelf life, and end product usage. A non-thermal method, effective on one food product may not apply to other food products. The degree of treatment and exposure time will vary based on the matrix of the product.

1.6 Summary

- Three fundamental approaches for food preservation are inactivation, inhibition, and restriction of microbial population or growth.
- Treatment of food at a higher intensity of temperature and time leads to an increased microbial safety but with a compromised nutritional quality.
- Non-thermal treatment ensures the microbial safety and extension of shelf life of the food and exerts a minimal impact on the nutritional and sensory properties of food.
- Most popularly used non-thermal techniques are HPP, PEF, PL, US, CP treatments, irradiation, OMF, and ozone treatment.
- Most of the non-thermal technique acts primarily on the inactivation of microorganisms.
- The HPP, US, and OMF treatments can be employed in packaged food, thus eliminating the possibility of post-processing contamination.
- The non-thermal process intensity for a high-value food product can be designed while satisfying the demand of three stakeholders, viz. manufacturer (microbial safety), retailer (microbial and enzymatic stability), and consumers (nutrient retention).
- Using a non-thermal technique for a food product depends upon the equipment cost, scale, and economy of production, type, and the desired shelf life.
- HPP and PEF are the most industrially accepted non-thermal treatments for food products.
- The use of radioactive sources for food preservation is still a concern in some countries.

1.7 Multiple Choice Questions

1. Which of the following is not a fundamental approach for food preservation?
   a. Restriction
   b. Inactivation
   c. Homogenization
   d. Inhibition

2. Keeping milk at a low temperature is an example of _____.
   a. Inactivation of the microbial population during cooling
   b. Restriction of microbial invasion
   c. Retention of the flavor of the milk
   d. Inhibition of microbial growth

3. Which of the following is not a non-thermal food preservation technique?
   a. Power ultrasound
   b. Pulsed light
   c. Dehydration
   d. Pulsed electric field

4. Which one of the following is the principal mechanism of action of pulsed electric field treatment?
   a. Electroporation
   b. Cavitation
   c. Oxidation of proteins
   d. Contractions of membranes

5. What is the principal mechanism of action of ozonization?
   a. Oxidation of proteins
   b. Pascalization
   c. Cavitation
   d. Permeabilization

6. Which one of the following non-thermal techniques are mainly used for surface sanitation?
   a. High-pressure processing
   b. Pulsed electric field
   c. Oscillating magnetic field
   d. Cold plasma

7. High-pressure processing is also known as _____.
   a. Pascalization
   b. Radurization
   c. Tyndallization
   d. Appertization

8. Which of the following is not the principal objective of non-thermal food processing?
   a. To ensure microbial safety of the food products
   b. To extend the shelf life of food products
   c. To impart a particular flavor to the food product
   d. To retain the bioactive or phytochemicals in the food product

9. Which one of the following non-thermal techniques is not employed on a packaged food product?
   a. High-pressure processing
   b. Pulsed light treatment
   c. Irradiation
   d. Oscillating magnetic field

10. For food preservation using gamma rays, the approved source for radiation is _____.
   a. Potassium - 40
   b. Cobalt - 60
   c. Thorium - 228
   d. Uranium - 234

1.8 Short Answer Type Questions

a. What are the basic approaches for shelf-life extension of food?
b. List some non-thermal techniques and corresponding forms of energy used to induce stress to microorganisms.
c. Why does the non-thermal treatment is favored for high-value food products?
d. What are the limitations of non-thermally treated food products?
e. What are the factors influencing the effectiveness of any non-thermal treatment?

1.9 Descriptive Questions

a. What are the main advantages of non-thermal processing over thermal processing?

b. Describe the difference in principle action on microorganisms in food during thermal and non-thermal treatments.

c. Discuss the strategy to design the intensity of a non-thermal treatment for a high-value food product.

d. Highlight the principle of actions on the microorganisms of the non-thermal treatments applied mainly to liquid food products.

e. Describe the effects of non-thermal processing on spoilage enzymes and nutrients in food product.

Suggested Readings

Chauhan, O. P. (2019). Non-thermal Processing of Foods, CRC Press-Taylor & Francis.

Cullen, P. J., Tiwari, B. K., & Valdramidis, V. P. (2012). *Novel Thermal and Non-thermal Technologies for Fluid Foods*, Elsevier Academic Press.

Ortega-Rivas, E. (2012). *Non-thermal Food Engineering Operations*, Springer US.

Sun, Da-Wen. (2005). *Emerging Technologies for Food Processing*, Elsevier Academic Press.

Van Impe, J., Smet, C., Tiwari, B., Greiner, R., Ojha, S., Stulić, V., ... & Režek Jambrak, A. (2018). State of the art of non-thermal and thermal processing for inactivation of microorganisms. *Journal of Applied Microbiology*, *125*(1), 16–35.

Zhang, H. Q., Barbosa-Cánovas, G. V., Bala Balasubramaniam, V. M., Dunne, C. P., Farkas, D. F., & Yuan, J. T. C. (2011). *Non-thermal Processing Technologies for Food*, John Wiley & Sons.

Answers for MCQs (sec. 1.7)

1	2	3	4	5	6	7	8	9	10
c	d	c	a	a	d	a	c	b	b

2

HIGH-PRESSURE PROCESSING

2.1 Principle

High-pressure processing (HPP) is a non-thermal technique in which the food material (packaged or without packaging) is subjected to a hydrostatic pressure ranging from 50 to 1000 MPa. The microorganisms are inactivated by the stress from this huge pressure magnitude. Realizing the magnitude of pressure, the atmospheric pressure is 1 atm = 101.325 kPa = 0.1 MPa ≈ 1 bar. That means the pressure applied in HPP is 500–10,000 bar (500–10,000 times the atmospheric pressure). This HPP technique is also known as ultra-HPP, high hydrostatic pressure processing, pascalization, or hyperbaric pressure processing. Fundamentally, HPP works on Le Chatelier's principle and isostatic rule. Le Chatelier's principle describes; with the increase in applied pressure, positive or negative change in the total volume shifts the equilibrium toward bond breaking or bond formation, respectively. The application of high pressure facilitates volume reduction, thus favoring the chain reactions that lead to new bond formation.

Consequently, high pressure can disrupt large molecules of microbial cell structures, such as enzymes, proteins, lipids, and cell membranes, leaving small molecules, such as vitamins and flavor components, unaffected. The effects of high pressure are uniform as its distribution follows the isostatic rule, which is nearly instantaneous throughout the food irrespective of its geometry and equipment size. For instance, the grape will not be distorted or torn after the pressure cycle if we pressurize one whole grape dipped within the pressurizing medium.

2.2 Operation

In the case of HPP, the generation of pressure in the form of hydrostatic pressure is crucial. Hydrostatic pressure means that the food will stay within the medium, and the desired magnitude of pressure

DOI: 10.1201/9781003199809-2   **17**

will be generated by compressing a liquid medium, called pressure-transmitting medium (PTM). The operation for an HPP system is accomplished either in batch mode or in semicontinuous mode. A continuous HPP system has not been explored yet.

2.2.1 Batch Mode

A schematic describing the sequential batch operation of HPP has been presented in **Fig. 2.1**. In the batch HPP process, the typical steps are (1) loading of products within the pressure vessel; (2) filling up the vessel by PTM; (3) compression of PTM and holding the product at desired pressure for a specific duration; (4) release of pressure and unloading of the products. The fluid food material is preferably packed within a flexible packaging material at product loading. In some cases, solid foods are placed within the vessel without any packaging. While processing, initially, it takes time to reach a specific pressure value known as the come-up time for pressure generation. The rate of pressure build-up in the vessel is termed ramp rate. After reaching the desired pressure, our countdown for treatment time starts, and the product is left under constant pressure; this is the holding or dwelling time. Once the specific holding time is over, decompression of the

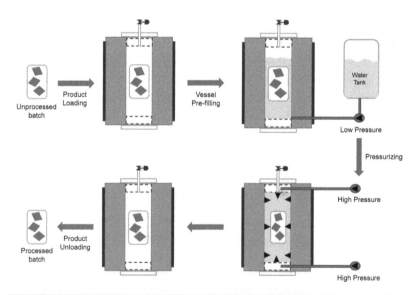

Figure 2.1   Schematic showing the sequential operations in a batch-type high-pressure processing unit.

vessel is done. Therefore, a batch HPP cycle consists of three segments: (1) compression, (2) holding, and (3) decompression. The cycle time is the cumulative time required for all those three segments. From an industrial facet, maximum pressure of 600 MPa is used for the HPP of food.

2.2.2 Semicontinuous Mode

In the semicontinuous mode, more than one pressure vessel is working simultaneously (**Fig. 2.2**). Typically, liquid foods are processed through this mode, and after depressurization or decompression, an aseptic environment is maintained during packaging. If the pumping of the product is going on into the first vessel, the sequential vessel will be associated with either pressure treatment (holding) or unloading.

In brief, a free piston is there within the vessel, and it moves up while the vessel is filled with the product. Once the vessel is filled in, and the feed valve is closed, the PTM is pumped through high-pressure pumps to compress the free piston from the top. In this way, the fluid inside the vessel gets compressed, and, after the desired holding time, the pressure is released. Further, the piston moves toward the discharge port, and the high-pressure treated fluid is collected from the outlet. In Japan, the industry uses the semicontinuous HPP system with a capacity of 600 L/h operating at 400 MPa.

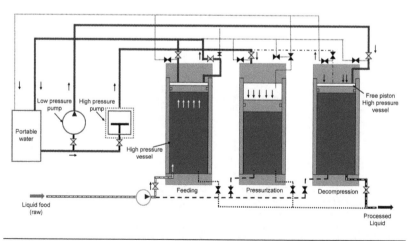

Figure 2.2 Schematic showing the operations in a multi-vessel semi-continuous HPP unit. (Redrawn from Singh, R.P. (2001) Technical Elements of New and Emerging Non-thermal Food Technologies, FAO bulletin)

2.3 Components of High-Pressure Processing System

A batch-type HPP system has at least three major components: (1) pressure vessel, (2) pressure generation and release assembly, and (3) controller unit (PLC module). A laboratory model batch-type HPP system has been shown in **Fig. 2.3**. A programmable logic controller (PLC) decides the operating parameters, such as rate of compression, decompression, holding time, target pressure, and temperature. The high-pressure vessel is kept inside the vessel module, and the service module consists of assembly for compression, decompression, and temperature control.

2.3.1 Pressure Vessel

The pressure vessel is the crucial component of the entire HPP system. A cylindrical shape is preferred as a pressure vessel. The low-alloy steel with high-tensile strength is used to withstand the pressure. Depending on the maximum pressure during the process, the number of cycles, working volume, and the wall thickness of the vessel is designed.

The forged monolithic wall material is used for low-pressure operations like 200–400 MPa, whereas, for 400 MPa or above, wire-winding is employed in the vessel construction. Various mechanisms are used regarding the top and bottom closure of the vessel, such as multi-threaded and yoke-driven mechanisms. The yoke is a heavy metal assembly that fixes the top and bottom seals at a desired position of the pressure vessel.

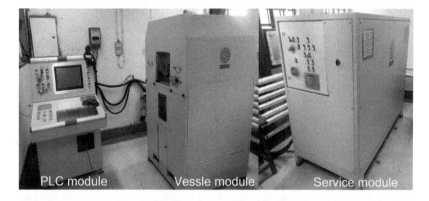

Figure 2.3 A laboratory model batch type high-pressure processing system installed at Agricultural and Food Engineering Department, Indian Institute of Technology, Kharagpur, India.

2.3.2 Pressure Generation System

The massive magnitude of pressure, such as 600–1000 MPa, is generated within the PTM in two ways: direct and indirect pressurization (**Fig. 2.4**). Pressure build-up using piston movement is referred to as direct pressurization. The piston directly compresses the PTM inside the vessel. On the other hand, indirect pressurization includes a pump and/or intensifier for PTM displacement and temperature control. In the case of indirect pressurization, two liquids, viz. oil (intensifier side) and PTM (pressure vessel side), are involved across the pump piston. The intensifier pressurizes the oil to purge the PTM inside the vessel to generate the desired pressure level. Industrially, indirect pressurization is preferred over direct one. This eliminates the possibility of contaminating the PTM by metal particles from the corroded piston, which may block the capillary.

2.3.3 Pressure-Transmitting Medium

The generated pressure is transferred to the food material through the PTM. This helps create uniform pressure (isostatic condition) all around the product. The compressibility of the PTM is crucial. For instance, water is 20,000 times less compressible than air, making the former a suitable PTM. The solvent to be used as PTM should

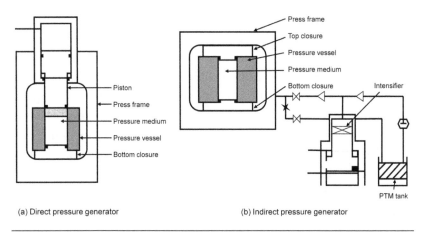

(a) Direct pressure generator   (b) Indirect pressure generator

Figure 2.4   Two different methods for a high-pressure generation. (Redrawn from Kanda, T. (1990). Recent trend of industrial high-pressure equipment and its application to food processing. In Hayashi, R. (Ed.), Pressure-Processed Food Research and Development. Japan. p. 341)

be food-grade. In most cases, pure water is used as PTM. For high-or low-temperature operations, solute such as mono-propylene glycol (MPG) or isopropyl alcohol (IPA) is mixed with water to modulate the freezing point.

2.3.4 Temperature Controller

The temperature inside the vessel is controlled by the heating and cooling jacket surrounding the vessel (**Fig. 2.1**). The temperature inside the jacket is maintained so that the heat is being conducted through the wall material and raises the temperature of the PTM. Internal heating or cooling inside the vessel is also employed in some cases. Pressure resistant thermocouples (J- or K-type) are placed at the desired places to monitor and maintain the temperature profile.

2.3.5 Heat of Compression

From a thermodynamic point of view, compression of PTM results in adiabatic heating, which increases the temperature of the medium. The extent of temperature rise depends on the specific heat of compression of the PTM (Eq. 2.1).

$$\left(\frac{\partial T}{\partial P}\right)_{Adiabatic} = \frac{\beta T}{\rho C_p} \tag{2.1}$$

In Eq. 2.1, T and P represent temperature (in K) and pressure (MPa), respectively; ρ, C_p, and β stand for density (kg·m^{-3}), specific heat capacity (J·kg^{-1}·K^{-1}), and compressibility of the PTM, respectively. For water as the PTM, the corresponding temperature rise ranges between 2.5 and 3°C for every 100 MPa elevation in pressure level. That means keeping water as PTM; there will be a temperature rise of 15–18°C, if the target pressure is 600 MPa. Similarly, for fats and oils as PTM, the magnitude of temperature rise varies from 6 to 8°C for every 100 MPa. From an industrial facet, 600 MPa is used as the upper limit for an economic HPP unit operating with 600–1200 L batch capacity. On a similar note, while decompression, there will be adiabatic cooling of PTM, leading to a similar rate of temperature drop. The compressibility (β) of the medium can be connected to a change in volume as a gradient of pressure at constant temperature (Eq. 2.2).

$$\beta = -\frac{1}{V}\left(\frac{\partial V}{\partial P}\right)_T \qquad (2.2)$$

In Eq. 2.2, V represents the volume (m³) of the PTM. Within the HPP cycle, if the compression follows the equilibrium path, the magnitude of work (W) during rapid pressurization (also reduction in volume) can be correlated as Eq. 2.3.

$$W = \int PV\beta\, dP - \frac{1}{V}\left(\frac{\partial V}{\partial P}\right)_T \qquad (2.3)$$

Eq. 2.3 states that if the pressure is fixed (during holding time), the corresponding work input is zero. This is one of the crucial advantages of HPP, which says that no operating work input is required during dwell time or holding time of the cycle.

2.3.6 Processing Time

Any high-pressure cycle consists of compression, holding, and decompression. The compression rate, also called ramp rate, defines the time required to reach the target pressure. A representative pressure-temperature profile for an HPP cycle has been shown in **Fig. 2.5**. As can be seen from the figure, the ramp rate is about 300 MPa/min, which leads to compression or come-up time just above 2 min. Due to adiabatic compression, the temperature of the PTM also increases during this time and reaches a maximum value at the end of the come-up time. Once the target pressure is reached, the compression ends, and the HPP system is kept hold with both the top and bottom seals closed. This period is called holding time. The pressure remains almost constant during this period, and the temperature is also maintained through the heating-cooling jacket. Once the desired holding time is completed, the valves are opened, and the pressure inside the vessel is released. This depressurization is instantaneous (within 10–20 seconds), and the corresponding period is called decompression time. The temperature of the PTM decreases rapidly due to adiabatic heat loss from the pressure drop. The cumulative time required starting

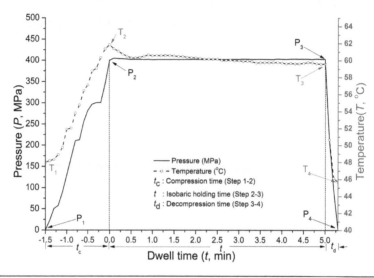

Figure 2.5 The representative HPP cycle's pressure and temperature profile (400 MPa/60°C/5 min). (Chakraborty et al., 2019)

from compression (from atmospheric pressure of 0.1 MPa to target pressure like 600 MPa), holding, and decompression (from 600 MPa to 0.1 MPa) is called the cycle time for HPP.

2.3.7 Suitable Packaging Materials

The packaging material for HPP needs to be flexible, which can withstand the maximum applied pressure, compression, and decompression steps. Plastic films are preferred as suitable packaging material for HPP, and besides, the presence of air inside the package is not recommended. To name a few, ethylene-vinyl alcohol copolymer (EVOH) and polyvinyl alcohol (PVOH) are used for HPP. Eventually, these are not acceptable for high-temperature processing. Obviously, due to limited or no flexibility, the metal cans and glasses are avoided to be processed inside the HPP vessel.

2.4 Mechanism of Quality Changes in Food

2.4.1 Mechanism Microbial Inactivation

The application of high pressure, targets a single site of the microbial cell but it hits at multiple sites. Microbial inactivation through HPP is achieved mainly by targeting the cell membrane. When food is subjected to high pressure, the hydrostatic medium exhibits the compression of microbes present in the food. This alters the surface morphology and permeability of the cell membrane resulting in the leakage of intracellular constituents through the perforated cell membrane. It is the most common reason for microbial death by high-pressure treatment. Sometimes applied pressure is insufficient to permeabilize the cell membrane or restore pressure release. Moreover, the fluidity of the cell membrane also affects processing efficacy. Microorganisms with rigid cell walls are susceptible to high pressure. Oppositely high fluidity in cell membrane protects against HPP.

Compression due to HPP appears to affect microbial inactivation by changing or reforming the proteins responsible for replication and metabolism. Moreover, high-pressure application alters the hydrogen and ionic bonds that hold the proteins in their original forms. Besides the type of microorganism, composition of suspension media or food, pressure level, and treatment time, the critical parameters for high pressure-induced microbial inactivation also depend on pH, water activity (a_w), and the treatment temperature.

2.4.2 Mechanism of Inactivating Spoilage Enzymes

An enzyme is a special molecule that is responsible for several biochemical reactions because of the uniqueness of its catalytic power (the ability of an enzyme to accelerate a chemical reaction) and maintenance of specificity of the active site (three-dimensional hydrophobic cleft bonded with a substrate). In HPP conditions, the mechanism of enzyme inactivation can be discussed similarly to protein denaturation or alteration. Like proteins, enzyme conformation is characterized in primary, secondary, tertiary, and quaternary structures.

The application of high pressure on enzymes exhibits reversible or irreversible and partial or complete unfolding of the native structure, resulting in the alteration of the structure of its active site, which eventually leads to a change in enzyme activity. Due to several covalent bonding, the primary structure doesn't affect much due to pressurization. Still, more complex structures like tertiary and quaternary structures alter the hydrogen bonds, hydrophobic (reject to mix with water), and electrostatic (stationary electric charges) interactions. Moreover, high pressure might assist the formation of intermolecular disulfide bond formations due to oxidation of sulfhydryl to disulfide groups, which stabilize the enzyme. This addition of osmolytes (amino acids, sugars, and polyols) to the medium also changes the enzyme activity by altering the hydration capacity of the enzyme.

2.4.3 Mechanism of Nutritional Quality Change

The complex and large molecular weight compounds, such as proteins, enzymes, nucleic acid, and polysaccharides, along with low-molecular-weight food components, such as flavoring agents, pigments, vitamins, bioactive compounds, etc., describe the nutritional quality of food. The compression of volume induced due to pressurization is the only main reason for changes in molecules' interactions, bonds, and structure. The alteration in hydrogen bond and hydrophobic interactions in proteins, nucleic acid, and polysaccharides might be possible as volume compaction can occur. In contrast, the covalent bonds (an interaction that involves the sharing of electron pairs between atoms) in the primary structure of low molecular compounds are not affected by pressurization. Hence, they remain unaffected; moreover, they are better preserved by HPP than by thermal treatment.

2.5 Critical Parameters for Process Design

2.5.1 Type of Product

HPP is capable of processing almost all types of material. This technology has been applied to various food products, such as fruits, fruit juices, milk, milk products, meat, smoothies, ham, salsa, rice products, fish, etc. Solid foods are best suited for batch-type HPP. The food matrix and pH strongly influence the efficacy of HPP to inactivate microorganisms. On the other hand, the product's water activity, total soluble solids (TSS), amino acids, sugars, and polyols can significantly impact enzyme inactivation.

2.5.2 Target Microorganism or Enzyme

Microorganisms and enzymes may have variable resistance toward HPP treatment. The susceptibility of gram-negative bacteria (*Escherichia coli* and *Pseudomonas* spp.) is the highest toward HPP. Gram-positive (such as *Listeria* and *Bacillus* spp.) have moderate susceptibility, and fungal spores (such as spores of *Saccharomyces cerevisiae*) or bacterial spores (such as spores of *Bacillus* spp. and *Clostridium* spp.) are the most resistant. Bacterial spores require a pressure higher than 1200 MPa to inactivate them. Different enzymes or the same enzyme from different sources may show different susceptibility. For instance, in a study of HPP (600 MPa pressure for 120 min) of mango pulp, the researchers found the order of susceptibility as; polyphenol oxidase (PPO) > pectin methylesterase (PME) > peroxidase (POD). But this trend may not hold for mango pulp of some other variety or a completely different fruit. The initial microbial load or concentrations of enzymes in the product can also be an important factor during the inactivation process.

2.5.3 Equipment and Mode of Operation

Naturally, applying higher pressure and exposure time can attain better and faster inactivation of microorganisms and spoilage enzymes. Even though HPP is intended to be a non-thermal process, the mild temperature rise may occur due to the intrinsic property of high pressure, which may synergistically enhance the inactivation of microorganisms and spoilage enzymes. The inactivation rate constant (k) of microorganisms or enzymes can be related to the change in temperature (T) by Arrhenius's equation (Eq. 2.4, at a fixed pressure, P_{ref}) and change in pressure (P) by Eyring's equation (Eq. 2.5, at a fixed temperature, T_{ref}) during HPP.

$$\ln(k) = \ln(k_{ref}) + \frac{E_a}{R}\left(\frac{1}{T_{ref}} - \frac{1}{T}\right) \tag{2.4}$$

$$\ln(k) = \ln(k_{ref}) + \frac{V_a}{RT}\left(P_{ref} - P\right) \tag{2.5}$$

Here k_{ref} is the reference rate constant, $\ln(k_{ref})$ acts as an intercept for both the equations, E_a is the activation energy, V_a is the activation volume, T_{ref} is the reference temperature, P_{ref} is the reference pressure, and R is the universal gas constant (8.314 J/(K·mol)). The Eq. 2.4 and Eq. 2.5 are secondary models applied after calculating the k values from primary kinetic models. The primary models can be of any order (1st to n^{th}) depending on the target microorganism or enzyme and other internal or external factors. Generally, the primary kinetic models, such as first-order type (Eq. 2.6), express k as a function of time (t) and survival fraction (s) at fixed temperature and pressure. As no variable term for temperature or pressure exists in its equation, kinetic models are only applicable during a continuous holding period when temperature and pressure remain constant.

$$s = \log\left(\frac{N}{N_i}\right) = k \cdot t \tag{2.6}$$

In Eq. 2.6 N and N_i are the current and initial microbial counts. To consider the inactivation that occurred during the dynamic phases of come-up and fall-down time (together can be called as pulse effect) for certain pressure and/or temperature, a different treatment of 1 s holding time is employed. It can give us information regarding the inactivation happening only due to the come-up and decompression (falling-down) time as the holding time of 1 s is too negligible to have any extra impact (**Fig. 2.6**). Eq. 2.7 explains the relation of pulse effect (P_E), in which N_o is the microbial count after 1 s pulse treatment.

$$P_E = -\log\left(\frac{N_o}{N_i}\right) \tag{2.7}$$

The batch mode is preferred for in-pack HPP, i.e., treating a packed food product that eliminates the chance of post-processing or cross-contamination. Semicontinuous and bulk modes are applicable for liquid and pumpable fluids only, and the product is packaged afterward.

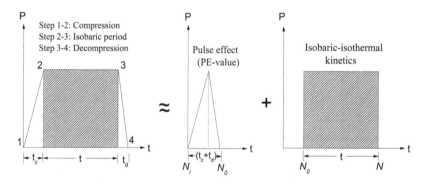

Figure 2.6   Schematic of segregating the combined effect of compression and decompression from the lethality of the entire HPP cycle. (Chakraborty et al., 2014, 2015)

Bulk processing is about 25% more efficient than a semicontinuous one, but it demands an aseptic design and arrangement of the HPP system.

2.6 Practical Example

We will consider one HPP-based research work as a practical example for better understanding. In a study regarding HPP of pineapple puree where the effect of processing has been recorded for microorganisms, including aerobic mesophiles (AM) and yeast and mold (YM) (**Table 2.1**). The microbial counts were quantified as CFU (colony forming unit) per g of puree. In addition, spoilage enzymes such as PPO activity, total color change (ΔE^*), antioxidant capacity, and vitamin C content between two equivalent thermal and HPP treated samples were estimated. Equivalent condition refers to the inactivation of natural microflora, such as AM and YM, besides 90% inactivation of PPO activity. The corresponding HPP treatment condition was 600 MPa/70°C/20 min, whereas the thermal treatment condition was 0.1 MPa (atmospheric pressure)/95°C/12 min. While comparing the other quality attributes between these two samples with respect to the untreated sample, the thermally treated sample ($\Delta E^* = 9.46$) appeared to be browner than HPP treated sample ($\Delta E^* = 6.92$). The HPP treated sample retained 53% and 55% more vitamin C and antioxidant capacity, respectively than the thermally treated samples. Thus, HPP produced a product with superior phytochemical content than the equivalent thermal treatment.

Table 2.1   Comparison in Quality Attributes between Equivalent Thermal and HPP Treated Pineapple Puree (Chakraborty et al. 2015)

	SAMPLE TREATED AT MPa/°C/MIN		
QUALITY ATTRIBUTES	UNTREATED	600/70/20	0.1/95/12
Total color change (ΔE^*)	-	6.92 ± 0.43	9.46 ± 0.77
Antioxidant capacity (mg GAEAC/100 g sample)	12.4 ± 2.7	8.7 ± 0.8	1.9 ± 0.7
Ascorbic acid (mg/100 g sample)	54.0 ± 5.6	41.4 ± 1.3	12.5 ± 0.7
Relative activity of PPO (%)	100	8.1 ± 1.2	2.7 ± 1.4
Aerobic mesophilic count (Log cfu/g)	6.12 ± 0.76	<DL	<DL
Yeasts and molds count (Log cfu/g)	5.76 ± 1.03	<DL	<DL

The total color change is calculated taking untreated samples as a reference; DL, detection limit; GAE, gallic acid equivalent; GAEAC, gallic acid equivalent antioxidant capacity.

2.7 Potential and Challenges

Industrially HPP is being applied for a wide range of food products, including fruit and vegetable products and beverages, dairy products, seafood, meat products, ready-to-eat meals, infant foods, etc. In most cases, the large-scale HPP treatment is employed at ambient temperature, keeping the pressure within 400–600 MPa, and treatment time ranges up to 15 minutes. However, the treatment conditions vary for each product, depending on the composition and target lethality. HPP reduces microbial load and inactivates the enzymes, ensuring safe, high quality, and shelf-stable food products. HPP induces minimal or no loss of vitamins and other essential nutrients, unlike thermal processing, along with replicating the sensory properties of fresh produce. The intensity of HPP treatment is independent of mass, and it enables instant transmittance of pressure throughout the system, irrespective of size and geometry, and allowing, uniform treatment.

Pasteurization can be achieved by high-pressure treatment even under low temperatures, which retains the nutritional qualities of foods. HPP can be used to retard the browning reaction in foods. The condensation in a Maillard reaction does not show any acceleration by high-pressure treatment due to the suppression of free, stable radicals derived from melanoidin. Protein gels induced by high-pressure treatment are glossy and transparent because of the rearrangement of water molecules surrounding the amino acid residues in a denatured state. On the other hand, the capital cost of HPP is very high, making it difficult for a commercial scale-up. Implementation of comprehensive quality assurance of the high-pressure-treated product in microbial safety finally leads to higher processing costs. The equipment of the HPP system is very complex and requires extremely high precision in its construction, utilization, and maintenance. Separate expert or trained personnel is necessary for handling the equipment.

2.8 Summary

- High-pressure processing is a non-thermal process in which hydrostatic pressure acts as the stress to inactivate microorganisms and enzymes.
- Hydrostatic pressure distributes inside the pressure vessel through an isostatic way.

- High-pressure does not affect the covalent bonds, thus retains the maximum phytochemical and nutrients inside the HPP treated food.
- A flexible pouch is preferred for HPP treatment.
- The post-processing contamination is avoided by packing the sample before HPP.
- High-pressure can be generated by direct or indirect pressurization method.
- During the pressure holding time, there is no work input from the pump and piston.
- The HPP batch cycle consists of compression, holding, and decompression zones.
- Pulse effect is the cumulative lethality within the HPP sample with a single second of holding time.
- A high-equipment cost is one of the limitations before its wide application.

2.9 Solved Numerical

1) The total plate count obtained in a sample of pineapple chunks after several pressure and temperature combinations has been given below. The inactivation trend followed a Weibull distribution with respect to time, with a scale parameter of b and a shape parameter of n, respectively. The logarithmic term of the calculated scale parameter (b_c) at an average value of n (n_{avg}) followed the linear relationship with the applied pressure and a slope of reciprocal Z_p. Find out Z_p and pulse effect (P_E) values for 200 MPa and 400 MPa at 40°C.

PRESSURE (MPa)	TEMPERATURE (°C)	TIME	COUNT (N in Log CFU·g^{-1})
0.1 (atm pressure)	25	Untreated	6.890
200	40	1 s	6.091
200	40	5 min	5.434
200	40	10 min	4.473
400	40	1 s	5.791
400	40	5 min	4.370
400	40	10 min	2.602

Solution: Count of 1 s treated sample has been taken as reference, so, N_o at 200 MPa = 6.091 log cfu·g^{-1} and N_o at 400 MPa = 5.791 log cfu·g^{-1}

$$\log(N_t/N_o) = -b \cdot t^n \qquad \Longrightarrow (\log N_t - \log N_o) = -b \cdot t^n$$
$$\Longrightarrow (\log N_o - \log N_t) = b \cdot t^n$$

Estimating n, using equation: $\log (\log N_o - \log N_t) = \log b + n \log t$

a. For 200 MPa at 40°C

Time: 10 min $\qquad\qquad$ $\log (6.091 - 4.473) = \cancel{\log b} + n \log 10$

Time: 5 min $\qquad\qquad$ $-\{\log (6.091 - 5.434) = \cancel{\log b} + n \log 5\}$

$\overline{\qquad\qquad \log (1.618) - \log (0.657) = n\, (\log 10 - \log 5)}$

$\Longrightarrow 0.391 = n(1 - 0.699) \qquad \Longrightarrow n = 1.300$

b. For 400 MPa at 40°C

Time: 10 min $\qquad\qquad$ $\log (5.791 - 2.602) = \cancel{\log b} + n \log 10$

Time: 5 min $\qquad\qquad$ $-\{\log (5.791 - 4.370) = \cancel{\log b} + n \log 5\}$

$\overline{\qquad\qquad \log (3.189) - \log (1.421) = n\, (\log 10 - \log 5)}$

$\Longrightarrow 0.351 = n(1 - 0.699) \qquad \Longrightarrow n = 1.166$

Average shape factor, $n_{avg} = \frac{(1.300 + 1.166)}{2} = 1.233$

[*Note*: The n_{avg} value will remain constant for the entire domain of pressure 200–400 MPa, but scale parameter b_c will change with change in pressure.]

The scale parameter (b_c) can now be calculated from shape parameter (n_{avg}) using equation:

$$(\log N_o - \log N_t) = b \cdot t^n$$

a. For 200 MPa at 40°C

Time: 5 min $\qquad$ $6.091 - 5.434 = b_c \times (5)^{1.233} \qquad \Longrightarrow b_c = 0.0903$

Time: 10 min $\qquad$ $6.091 - 4.473 = b_c \times (10)^{1.233} \qquad \Longrightarrow b_c = 0.0946$

Average scale parameter at 200 MPa and 40°C,
$b_c = \frac{(0.0903 + 0.0946)}{2} = 0.09245$

b. For 400 MPa at 40°C

Time: 5 min $5.791 - 4.370 = b_c \times (5)^{1.233}$ $=> b_c = 0.1953$

Time: 10 min $5.791 - 2.602 = b_c \times (10)^{1.233}$ $=> b_c = 0.1865$

Average scale parameter at 400 MPa and 40°C,
$b_c = \frac{(0.1953 + 0.1865)}{2} = 0.1909$ min^{-1}

It has been given that the log of the calculated average scale parameter (log b_c) shares a linear relation with the pressure (P) having $1/Z_p$ as the slope. In other words, this relation follows the equation pattern of a straight line ($y = mx + c$).

$$\log b_c = (1/Z_p) \cdot P + C_1$$

400 MPa at 40°C:	$\log (0.1909) = (1/Z_p) \cdot 400 + C_1$
200 MPa at 40°C:	$-\{\log (0.09245) = (1/Z_p) \cdot 200 + C_1\}$

$$-0.719 - (-1.034) = (1/Z_p) \cdot (400 - 200)$$

$=> 0.315 = 200/Z_p$ $=> Z_p = 634.92$ MPa

Pulse effect calculation (at $t = 1$ s); $P_E = -\log (N_t/N_o) = -(\log N_t - \log N_i) = \log N_i - \log N_t$

P_E at 200 MPa at 40°C $= (6.89 - 6.091) = 0.799$ log cycles

P_E at 400 MPa at 40°C $= (6.89 - 5.791) = 1.099$ log cycles

2) Orange PME follows a first-order inactivation model in the range of 400–600 MPa. From an HP treatment at 18°C, inactivation rate constants of PME at 400, 500, and 600 MPa were 0.024, 0.107, and 0.229 min^{-1}, respectively. What is the average activation volume for PME in that experimental domain?

Solution: Temperature, $T = 18 + 273 = 291$ K and universal gas constant, $R = 8.314$ J·mol^{-1}·K^{-1}

Eyring's equation will be used: $\ln (k) = \ln (k_{ref}) + V_a/(RT) (P_{ref} - P)$

At 400 MPa pressure:

$\ln (0.024) = \ln (k_{ref}) + (V_a/(8.314 \times 291))(P_{ref} - 400)$

$=> -3.73 = \ln (k_{ref}) + (V_a/2419.374)(P_{ref} - 400)$

(Eq. A)

At 500 MPa pressure:

$$\ln(0.107) = \ln(k_{ref}) + (V_a/(8.314 \times 291))(P_{ref} - 500)$$
$$\Rightarrow -2.235 = \ln(k_{ref}) + (V_a/2419.374)(P_{ref} - 500) \quad \text{(Eq. B)}$$

At 600 MPa pressure:

$$\ln(0.227) = \ln(k_{ref}) + (V_a/(8.314 \times 291))(P_{ref} - 600)$$
$$\Rightarrow -1.483 = \ln(k_{ref}) + (V_a/2419.374)(P_{ref} - 600) \quad \text{(Eq. C)}$$

Operation: (A – B)

A: $-3.73 = \cancel{\ln(k_{ref})} + (V_a/2419.374)(P_{ref} - 400)$

B: $-\{-2.235 = \cancel{\ln(k_{ref})} + (V_a/2419.374)(P_{ref} - 500)\}$

$$\overline{-1.495 = (Va/2419.374)(500 - 400)}$$

$V_a = -36.17 \text{ mL} \cdot \text{mol}^{-1}$

Operation: (B – C)

B: $-2.235 = \cancel{\ln(k_{ref})} + (V_a/2419.374)(P_{ref} - 500)$

C: $-\{-1.483 = \cancel{\ln(k_{ref})} + (V_a/2419.374)(P_{ref} - 600)\}$

$$\overline{-0.752 = (V_a/2419.374)(600 - 500)}$$

$V_a = -18.19 \text{ mL} \cdot \text{mol}^{-1}$

Operation: (A – C)

A: $-3.73 = \cancel{\ln(k_{ref})} + (Va/2419.374)(P_{ref} - 400)$

C: $-\{-1.483 = \cancel{\ln(k_{ref})} + (Va/2419.374)(P_{ref} - 600)\}$

$$\overline{-2.247 = (Va/2419.374)(500 - 400)}$$

$Va = -27.182 \text{ mL} \cdot \text{mol}^{-1}$

Average activation volume, $Va = \frac{(-36.17) + (-18.19) + (-27.182)}{3} = -27.18 \text{ mL} \cdot \text{mol}^{-1}$

3) The inactivation of PME in peach pulp followed first-order kinetics within 100–800 MPa and 30–70°C. The rate constant (k in min^{-1}) was related to pressure (P in MPa) and temperature (T in K) by combining

Arrhenius-Eyring's empirical model, where $P_{ref} = 600$ MPa and $T_{ref} = 323$ K, respectively.

$$k = (k_{ref, P, T}) \times \exp[E_a/R(1/T_{ref} - 1/T) - V_a/RT(P - P_{ref})]$$

The activation energy (E_a), obtained from the Arrhenius model, at 100 and 800 MPa were 183 and 59 kJ·mol^{-1}, respectively, whereas the activation volume (V_a), obtained from Eyring's model, at 30 and 70°C were −39.3 and −9.9 mL·mol^{-1}, respectively. The E_a and V_a values were related to P and T, respectively, by the following equations:

$$E_a = E_{ao} \times \exp[-b(P - P_{ref})] \text{ and } V_a = V_{ao} + a(T - T_{ref})$$

Find the inactivation rates at 200 MPa/40°C and 700 MPa/60°C. (Given, $(k_{ref, P, T}) = 0.1212$ min^{-1} at 600 MPa/50°C.)

Solution: Calculating b and Ea_o using equation, $\ln(E_a) = \ln(E_{ao})^- b(P - P_{ref})$

$$\ln(183) = \ln(E_{ao}) - b(100 - 600)$$

$$\ln(59) = \ln(E_{ao}) - b(800 - 600)$$

Pressure 100 MPa: $\ln(183) = \ln(E_{ao}) - b(100 - 600)$

Pressure 800 MPa: $-\{\ln(59) = \ln(E_{ao}) - b(800 - 600)\}$

$$\overline{5.209 - 4.078 = b(500 + 200)}$$

=> $b = 0.00162$

Therefore, $E_{ao} = 81.41$ kJ mol^{-1}

Calculating a and V_{ao} using equation: $V_a = V_{a0} + a(T - T_{ref})$

Temperature: 30°C $-39.3 = V_{ao} + a(30 - 50)$

Temperature: 70°C $-\{-9.9 = V_{ao} + a(70 - 50)\}$

$$\overline{-29.4 = -a(20 + 20)}$$

=> $a = 0.735$

Therefore, $V_{ao} = -24.6$ mL·mol^{-1}

Estimating rate constant (k) using the combined Arrhenius-Eyring's empirical equation:

$$k = (k_{ref, P, T}) \times \exp[E_a/R(1/T_{ref} - 1/T) - V_a/RT(P - P_{ref})]$$

a. At 200 MPa at 40°C

Calculating 'E_a' from: $E_a = E_{ao} \times \exp[-b\,(P - P_{ref})]$

$$E_a = 81.41 \times \exp[-0.00162 \times (200 - 600)]$$

$$\Rightarrow E_a = 155.6 \text{ kJ} \cdot \text{mol}^{-1}$$

Calculating 'V_a' from: $V_a = V_{ao} + a\,(T - T_{ref})$

$$V_a = -24.6 + 0.735\,(40 - 50) \qquad \Rightarrow V_a = -31.95 \text{ mL} \cdot \text{mol}^{-1}$$

Now consider all the temperatures in Kelvin, $R = 0.008314$ kJ·mol^{-1}·k^{-1}, or $R = 8.314$ J·mol^{-1}·k^{-1}. Therefore, k at 200 MPa at 40°C

$$k = 0.1212 \times \exp\left[\frac{155.6}{0.008314} \times \left(\frac{1}{323} - \frac{1}{313} \right) \right.$$

$$\left. \times \frac{-31.95}{8.314 \times 313} \times (200 - 600) \right]$$

$$k = 1.4018 \times 10^{-4} \text{ min}^{-1}$$

b. At 700 MPa at 60°C

$$E_a = 81.41 \times \exp[-0.00162 \times (700 - 600)]$$

$$\Rightarrow E_a = 69.2 \text{ kJ} \cdot \text{mol}^{-1}$$

$$V_a = -24.6 + 0.735\,(60 - 50) \qquad \Rightarrow V_a = -17.25 \text{ mL} \cdot \text{mol}^{-1}$$

Therefore, k at 700 MPa at 60°C

$$k = 0.1212 \times \exp\left[\frac{69.2}{0.008314} \times \left(\frac{1}{323} - \frac{1}{333} \right) \right.$$

$$\left. \times \frac{-17.25}{8.314 \times 333} \times (700 - 600) \right]$$

$$k = 0.1212 \times \exp\left[\frac{69.2}{0.008314} \times \left(\frac{1}{323} - \frac{1}{333} \right) \right.$$

$$\left. - \frac{-17.25}{8.314 \times 333} \times (700 - 600) \right]$$

$$\Rightarrow k = 0.48997 \text{ min}^{-1}$$

2.10 Multiple Choice Questions

1. What type of enzyme structure is majorly responsible for inactivation by HPP?
   a. Tertiary and quaternary structures
   b. Secondary and tertiary structures
   c. Primary and secondary structures
   d. All the levels of structures

2. What is the expected heat of compression in the case of HPP of fats and oils?
   a. ~2–4°C/100 MPa
   b. ~4–6°C/100 MPa
   c. ~6–8°C/100 MPa
   d. ~8–10°C/100 MPa

3. The maximum product temperature at a specific processing pressure is independent of:
   a. Target pressure
   b. Compression rate
   c. Material compressibility
   d. Initial product temperature

4. Which type of microorganism was found to be the most baro-resistant?
   a. Gram-positive bacteria
   b. Gram-negative bacteria
   c. Yeast and mold
   d. All are equally susceptible

5. The magnitude of activation volume (V_a) present in the equation tends to _____ with increasing temperature, and a _____ magnitude of V_a indicates higher sensitivity of the target microorganism or enzyme toward ΔP.
   a. Increase, lower
   b. Increase, higher
   c. Decrease, lower
   d. Decrease, higher

6. Which one of the following is less preferred to be used as a PTM for HPP?
   a. Water
   b. Food grade oil
   c. 30% aqueous mono-propylene glycol
   d. Acetone

7. The zone within an HPP cycle having no work input is _____.
   a. Compression
   b. Holding
   c. Decompression
   d. Pumping the PTM inside the vessel before compression

8. The highest-pressure limit for an industrial HPP pasteurization technique is _____.
   a. 400 MPa
   b. 500 MPa
   c. 600 MPa
   d. 1000 MPa

9. Which one of the following does not associate with the working principle of HPP?
   a. Pascal's law of isostatic pressurization
   b. Dalton's law for partial pressure
   c. Le Chatelier's Principle
   d. Molecular reordering

10. Which one of the following does not associate with the design of the thickness of the pressure vessel wall?
    a. Working pressure
    b. Vessel diameter
    c. Number of cycles
    d. Compressibility of PTM

2.11 Short Answer Type Questions

a. Discuss the mechanism of high-pressure induced inactivation of microorganisms.
b. How does HPP affect the conformation of a fruit enzyme?

c. Continuous HPP equipment is not realistic to design. Comment.

d. How many zones are there for an HPP cycle? Discuss the work input for each zone.

e. What are the criteria for the packaging material to be employed in HPP?

2.12 Descriptive Questions

a. How does the high-pressure treatment retain the phytochemicals in the food?

b. Which are the fluid properties necessary to have to become pressure transmitting fluid?

c. Discuss the operation of a batch mode high-pressure processing for a food product?

d. Discuss the direct and indirect pressurization method for generating high hydrostatic pressure with a schematic.

e. Discuss the sustainability issues and challenges for a commercial HPP unit.

2.13 Numerical Problems

1) Within a high-pressure domain (100–600 MPa/30–70°C/ 1 s-30 min), the isobaric-isothermal inactivation rate constant of pineapple bromelain follows nth order reaction ($n = 1.2$). It also follows the combined Arrhenius-Eyring relationship where

$$E_a = E_{a0} + a_1(P - P_{ref}) + a_2(P - P_{ref})^2 \qquad V_a = V_0 + a_3(T - T_{ref})$$

Considering the midpoints of the P-T domain as reference values, the calculated activation energy (E_a) values at 200, 300, 500, and 600 MPa are 44.7, 31.4, 21.0, and 18.2 kJ·mol^{-1}, respectively. The activation volume (V_a) values at 50 and 70°C are −9.1 and −6.3 mL·mol^{-1}, respectively. Pulse effect (P_E) values are 0.06, 0.125, and 0.39 at 350 MPa/50°C, 400 MPa/60°C, and 600 MPa/70°C, respectively. If the inactivation rate of bromelain is 0.016 U^{n-1}·min^{-1} at 350 MPa/50°C, find out the time required for a 90% reduction in its initial activity at 400 MPa/60°C. Compare the same with first-order kinetics.

2) Within an HPP domain, the aerobic mesophilic (AM) count followed the Weibull distribution having a scale parameter of δ and shape parameter of β, respectively. The logarithmic term of the calculated scale parameter (δ_c) at β_{avg} followed the linear relationship with the applied pressure and a reciprocal slope of z_p. From the data below, find out z_p and pulse effect (P_E) values for both 250 MPa and 450 MPa at 40°C. Compare the z_p values with the log-linear model. In the untreated sample at 0.1 MPa at 25°C, AM count was 7 log cfu/g.

PRESSURE (MPa)	TEMPERATURE (°C)	AM COUNT (N, log (CFU/g) AFTER A HOLDING PERIOD OF:		
		1 sec	5 min	10 min
250	40	6.0	5.4	4.5
450	40	5.7	4.3	2.6

References

Chakraborty, S., Rao, P. S., & Mishra, H. N. (2015). Response surface optimization of process parameters and fuzzy analysis of sensory data of high pressure-temperature treated pineapple puree. *Journal of Food Science, 80*(8), E1763–E1775.

Suggested Readings

Chakraborty, S., Kaushik, N., & Mishra, H. N. (2018). High Pressure Processing of Fruit Juice and Puree. In H. N. Mishra (Ed.), *Food Products and Process Innovations.* New India Publishing Agency, New Delhi, India. Vol I, Chapter 11, pp. 305–344.

Chakraborty, S., Kaushik, N., Rao, P.S., & Mishra, H. N. (2014). High-pressure inactivation of enzymes: A review on its recent applications on fruit purees and juices. *Comprehensive Reviews in Food Science and Food Safety, 13*(4), 578–596.

Chakraborty, S., Swami Hulle, N. R., Jabeen, K., & Rao, P. S. (2016). Effect of Combined High Pressure-Temperature Treatments on Bioactive Compounds in Fruit Purees. In Jorge Moreno (Ed.), *Innovative Processing Technologies for Foods with Bioactive Compounds*, CRC Press, Boca Raton. Chapter 5, pp. 105–131.

Kanda, T. (1990). Recent Trend of Industrial High Pressure Equipment and Its Application to Food Processing. In R. Hayashi (Ed.), *Pressure-Processed Food Research and Development.* San-Ei Shuppan, Kyoto. p. 341.

Rao, P. S., Chakraborty, S., Kaushik, N., Kaur, B. P., & Swami, N. S. (2014). High Hydrostatic Pressure Processing of Food Materials. In J. K. Sahu (Ed.), *Introduction to Advanced Food Processing Technologies*. CRC Press, UK. Chapter 5, pp. 149–183.

Singh, R. P. (2001). Technical elements of new and emerging non-thermal food technologies, FAO bulletin.

Answers for MCQs (sec. 2.10)

1	2	3	4	5	6	7	8	9	10
a	c	b	c	d	d	b	c	b	d

3

PULSED ELECTRIC FIELD PROCESSING

3.1 Principles and Operation

In pulsed electric field (PEF) processing, the liquid or semisolid food product is kept between two electrodes enclosed in a treatment chamber. These electrodes allow the passage of high voltage pulses through the food material, which causes the product to experience high-energy electric fields only for a short duration (nanoseconds to microseconds). The gap between successive discharge and consecutive pulses during which no electrical energy is released into the product can vary from one millisecond to seconds. During PEF processing, the flow of electric current through the product can cause a temperature rise (ohmic or joule heating) but unlike the continuous flow of alternating current in ohmic heating technology (heating food by its resistance offered toward the flow of current), in PEF current flows through the product in the form of short pulses or bursts, which can allow far lower temperature rise compared to ohmic heating. PEF is intended to be used as a non-thermal processing technology. The strength or intensity of a constant uniform electric field, E (V/cm), experienced by the product is directly proportional to the voltage (V) applied across the electrodes and inversely proportional to the distance, d (cm) between the two electrodes (Eq. 3.1).

$$E = \frac{V}{d} \qquad (3.1)$$

Generally, PEF has been used for about 10–80 kV/cm electric field strength. The schematic of a typical continuous-type PEF system is shown in **Fig. 3.1**. Apart from the electrodes, treatment chamber (cell), and high voltage power supply, PEF consists of temporary energy-storing capacitors and a pulse generator to create the pulsation effect of discharge of the electrical energy. In the continuous

DOI: 10.1201/9781003199809-3 **43**

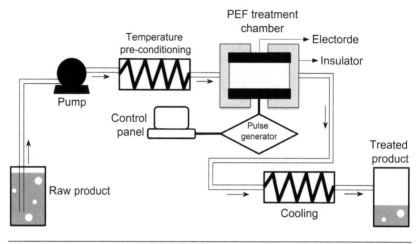

Figure 3.1   Continuous processing type pulse electric field system.

PEF assembly, raw material is pumped to a heat exchanger (preconditioning) followed by the PEF chamber. After the PEF treatment,
the product is cooled in a heat exchanger and stored in a tank. The
PEF chamber is placed inside an insulator. The parallel electrodes
are connected to a pulse generator regulated by a control panel.

3.2 Mechanism of Quality Changes in Food

3.2.1 Microbial Inactivation

Inactivation of microorganisms during PEF occurs due to a sharp
increase in membrane conductivity, permeability, and cell membrane
rupture. The detailed mechanism of PEF induced inactivation of
microorganisms has not yet been critically explored. However, two
significant theories were proposed: (1) dielectric rupture theory and
(2) electroporation theory (**Fig. 3.2**).

Dielectric rupture theory: When PEF is applied to a suspension
of microbial cells, accumulation of charge takes place both inside
and outside the cell membrane due to their difference in dielectric
constants. In turn, a transmembrane potential (TMP) is developed
across the membrane, which creates a situation similar to a capacitor. The intracellular materials and fluids act like a dielectric material
(for which electric polarization is possible, but current does not flow).
Food materials mainly possess a dielectric constant (ε, dimensionless)

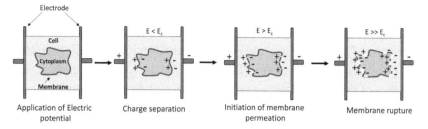

(a) Dielectric rupture theory

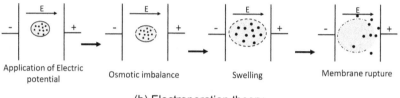

(b) Electroporation theory

Figure 3.2   Inactivation mechanism of microorganisms through; (a) dielectric rupture theory and (b) electroporation theory. (Adapted from Ortega-Rivas (2012) and Aguiló-Aguayo et al. (2012))

of about 60–80. In the direction of the strength of the applied electric field (E), TMP can be expressed as Eq. 3.2.

$$U_t \approx (0.75)d_c E \tag{3.2}$$

Where U_t (V) represents the TMP specifically in the direction of E (kV/mm) and d_c (μm) is the cell diameter. If microbial cells are of different shapes than spherical, a different expression can be used (Eq. 3.3).

$$U_t \approx \alpha d_c E \cos(\theta) \tag{3.3}$$

Where α is the shape parameter of the microbial cell ($\alpha = 0.75$ for spherical and 1 for rectangular cell) and θ is the angle between the electric field lines and the point on the membrane surface. The natural TMP that exits is about 10 mV. Under the influence of increasing electric fields formed by TMP, the cell membrane obtains opposite charges on the outside and inside via polarization. As the oppositely charged surfaces pull each other, the membrane experiences an electro-compression, resulting in reduced thickness that eventually causes membrane rupture. The cell membrane's natural viscoelasticity works against the electro-compression to close the pores or minute breakages. However, the force of electro-compression increases with increasing E (eventually crossing a critical electric field, E_C) that can easily overcome the viscoelastic membrane restoration causing the number of pores and their sizes to increase and damage the cell membrane irreversibly. Cytoplasm and cell organelles start to leak out due to abnormally high cell membrane permeability, which finally inactivates the microorganisms.

It has been reported that loss of membrane integrity and localized membrane breakdown occurs at a threshold TMP of approximately 1 V. Assuming a value of 1 μm cell diameter, an electric field of about 13.33 kV/cm is required to reach the threshold TMP of the cell membrane and overcome the reversible effects of cell's natural viscoelastic property (Eq. 3.2). Therefore, the applied electric field must be of high magnitude or long enough to have irreversible membrane damage involving the formation of pores; otherwise, the pores may get resealed again once the treatment ends.

Electroporation theory: Under the influence of the electric field, the cell membrane experiences both thermal and electrical effects. The cell membrane has finite permeability for ions. The lipids (lipid bilayer in the membrane) present possess a dipole nature, making the membrane very susceptible to the electric field (**Fig. 3.2**). The electric field instils conformational changes to the lipid present in the lipid bilayer and the protein molecules of the membrane by expanding the naturally existing pores or by generating new hydrophobic pores, eventually producing hydrophilic pores that are structurally more stable. Hydrophilic pores facilitate the flow of current, resulting in localized joule heating that promotes high-temperature rise for microseconds to milliseconds, which causes the thermal transformation of lipid bilayer from a rigid gel-like structure to a liquid crystalline structure. This perturbs the membrane's semipermeable characteristics, causing osmotic imbalance, swelling due to water influx, and, finally, cell lysis. The microbial cell consists of the cell membrane, protein channels, natural pores, and pumps. The closing and opening of these channels and pumps made of proteins are mainly dependent on TMP. The gating potential of these channels is around 50 mV which is significantly lower than the threshold TMP. Therefore, many of the voltage-sensitive protein channels will be opened under the impact of an applied electric field. After that, the current will enter the cell's body which is much more than the natural current during normal metabolic activities.

Consequently, the protein channels get denatured (disruption of the three-dimensional structure resulting in loss of native or natural characteristics and properties) by the action of joule heating (due to the flow of current inside the cell). It is coupled with the electrical alterations of functional sub-units, such as amino, carboxyl, hydroxyl, and sulfhydryl groups. In this way, the electroporation of the cell membrane affects both the lipid bilayer and protein channels leading to microbial inactivation.

3.2.2 Inactivation of Spoilage Enzymes

Enzymes are proteins that catalyze various biochemical reactions, and it is desirable to inactivate spoilage enzymes, such as polyphenoloxidase (PPO), peroxidase (POD), polygalacturonase (PG), pectin methylesterase (PME), and alkaline phosphatase (ALP), which deteriorates the food product, impairs organoleptic properties, and reduces their shelf life. The mechanisms for PEF-based inactivation of enzymes are still ambiguous. However, it has been hypothesized that localized momentary joule heating and electro-conformational changes can cause the enzyme proteins to unfold, aggregate, breakdown of covalent and disulfide bonds, and loss of secondary protein structural properties (such as β-sheet). Like enzymes, PEF may affect the proteins and amino acids present in the food product. Research on PEF showed that it could inactivate enzymes like PME and PG in certain fruit juice-milk beverages up to some extends, but not as good as thermal treatment.

3.2.3 Effect of Pulsed Electric Field on Nutritional Quality

The effect of PEF on nutritional qualities and the corresponding mechanism is not yet fully explored. Few research works showed that in terms of nutritional qualities of food, PEF treatment showed no significant impact on lipids, fatty acids, and carbohydrates of the food product. The bioactive components like polyphenols (known for their positive health benefits due to their antioxidant capacity) may undergo minor or no degradation depending on the food matrix and processing conditions. The high strength of the electric field and more prolonged treatment can reduce vitamin C (ascorbic acid), probably because of localized momentary joule heating and electro-conformational changes. Some literature reveals that PEF processing enhanced the diffusion of solids in juices, increased bioaccessibility of nutrients, and improved meat tenderness. Overall, PEF treatments have shown the potential to obtain food products with improved or better retained flavor, texture, and nutritional attributes retention.

3.3 Equipment and Critical Processing Factors

Generally, the PEF system comprises a high-voltage power source, energy-storing capacitor array, variable limiting resistors, and energy switch capable of discharging high energy from the capacitor bank to the product-containing chamber. The capacitor array gets charged by receiving the direct current from a rectified and amplified alternating current source. A laboratory model PEF assembly has been shown in **Fig. 3.3**. In addition to that, for a continuous system, a pump is used, and specific probes are installed to monitor voltage, current, and temperature. Followed by PEF treatment, the sample is preferably cooled by passing it through a cooling system. Several critical process parameters are associated with the PEF processing of food products.

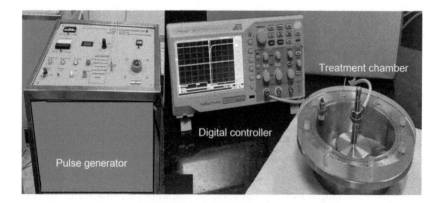

Figure 3.3 A laboratory model pulsed electric field (PEF) treatment assembly installed at the Food Engineering and Technology Department, Tezpur University, Assam, India.

3.3.1 Type of Product

PEF has been applied to various food materials, including fruit juices, milk, liquid egg, and meat products having different electrical properties. Food products contain polarizable dipolar molecules and ions responsible for the overall resistance and capacitance offered by the product, which accounts for the efficacy of PEF. The food product is a heterogeneous complex matrix that provides some resistance. The polarization action of the dipolar molecules provides capacitance (ability to store energy like a capacitor or condenser) and dielectric (electrical insulators capable of being polarized under electric field) properties. Each material possesses a specific resistance called the resistivity of that material, and conductivity is the opposite of resistivity. Products with better capacitance and dielectric properties are most suitable for PEF treatment; the higher the dielectric constant, the higher the capacity to hold the charge and ease the passage of the electric field. Products having better conductivity will further promote undesirable joule heating. A food material with high ions (like tomato juice) will show higher conductivity and poor resistivity. Liquid and semisolid products can be processed in a continuous or batch mode, but solid products can only be treated in a batch process. Other properties like lower pH can increase microbial inactivation by PEF. The product's initial temperature (preconditioning sample temperature) may also contribute to the overall impact of PEF on the product.

3.3.2 Target Microorganism

Different types of microorganisms show different susceptibility toward PEF. Research reveals that majorly yeasts (like *Saccharomyces cerevisiae*) are more sensitive than Gram-negative bacteria. Still, Gram-negative bacteria (like *Escherichia coli* and *Pseudomonas* spp.) are more sensitive than Gram-positive bacteria (like *Listeria* and *Bacillus* spp.). Yeast cells are comparatively larger than bacterial cells, and therefore exhibit lower breakdown/threshold TMP (Eq. 3.1), poor conductivity, and better capacitance. However, some researchers have also reported dissimilar or contradictory results, which could be due to differences in food product composition, variety of the product, microbial strains, and PEF processing conditions. Bacterial spores are resistant to PEF.

Different microorganisms show different rates of inactivation under PEF, which can be quantified using kinetic equations like; first-order and Weibull type kinetic models (Eq. 3.4).

$$\ln(s) = -\left(\frac{t}{\delta}\right)^{\beta} \quad \text{(Weibull)}$$

$$\ln(s) = -kt \quad \text{(First-order kinetics)}$$

(3.4)

Where,

- s = survival fraction; a ratio of current microbial count (N) to the initial microbial count (N_i)
- t = treatment time; a product of pulse width (on time, μs) and number of pulses
- k = specific inactivation rate constant for the microorganisms
- δ = scale parameter (s), reciprocal of rate constant (k)
- β = shape parameter

In addition, microbial cells at the logarithmic growth phase are generally more sensitive than their stationery and lag/death phase. In addition to that, a higher initial microbial load present in the product can reduce the efficacy of PEF treatment, which can be countered by increasing the severity of the process or reducing sample load or volume during processing.

3.3.3 Equipment and Mode of Operation

The treatment chamber of PEF has two electrodes clamped in the position by specific insulating material, forming an overall enclosure that provides containment for the food product. Different electrodes and chambers with varying configurations, such as parallel plates, rod-rod, rod-plate, wire-cylinder, collinear, and the coaxial (concentric) cylinder, can directly influence the PEF treatment (**Fig. 3.4**). Most of the conducted studies have used the parallel plate set-up, known for its uniform electric field strength and easier design and coaxial electrode design, which facilitated uniform and smooth product flow and is also suitable for continuous processing.

During PEF processing, different shapes of voltage waveforms can affect microbial inactivation efficiency. Different wave-forms exit,

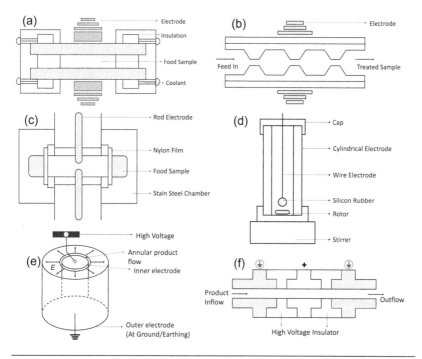

Figure 3.4 Design of different PEF devices; (a) static parallel plate system, (b) continuous parallel plate system (side view), (c) rod-rod electrode system, (d) wire-cylinder electrode system, (e) Coaxial chamber (horizontal cross-section), and (f) Collinear chamber.

such as square, oscillatory, and exponential form (**Fig. 3.5**). It has been reported that the square wave pulses are the most efficient, and the oscillatory decay pulses are the least efficient waveforms. However, bipolar pulses are more efficient than monopolar pulses (**Fig. 3.6**). The fluctuating stress provided by bipolar pulses leads to increased structural damage of microbial cell membrane and improves their susceptibility toward PEF, thus promoting better inactivation.

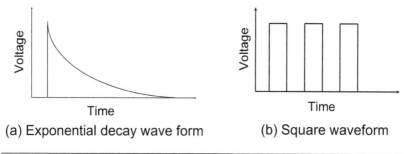

Figure 3.5 Exponential decay type (a) and square type (b) waveform of PEF.

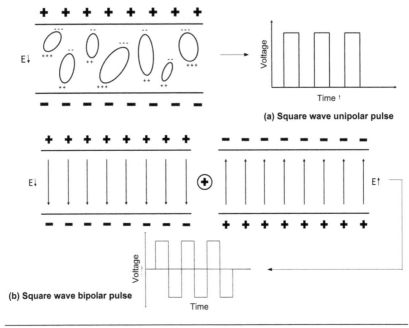

Figure 3.6   Unipolar (a) and bipolar (b) type square waveforms of PEF.

3.4 Practical Example

We will consider one of the PEF-based research work as a practical example for better understanding. Bansal et al. (2015) studied the effect of PEF on *Zygosaccharomyces bacilii* inoculated in amla (*Emblica officinalis*) juice compared with a specific thermal pasteurization condition. The authors have also studied the effect of processing on various other quality parameters. *Z. bacilii* is a spoilage yeast that can withstand low pH (<4.6) and chemical preservatives, and therefore can contaminate acidic food products, such as wine, pickle, certain fruit juices, and salad.

PEF treatment: The system was equipped with two stainless steel electrodes placed with a gap of 10 mm, and the juice was loaded in a 5-mL chamber. The optimum treatment condition includes monopolar rectangular pulse (1 μs), 26 kV/cm electric field, 1 Hz pulse frequency, and 500 μs treatment time. After treatment, the juice temperature did not exceed 41°C and was immediately stored at 4°C until further analysis.

Thermal pasteurization: A 20-mL juice was heated in a glass tube container kept in a hot water bath to reach a temperature of 90°C and

Table 3.1 Effect of PEF and Thermal Processing on *Z. bacilii* and Other Quality Parameters of Amla (Indian Gooseberry) Juice

TREATMENT	TSS (°Brix)	pH	*Z. bacilii* COUNT (Log CFU/mL)	VIT-C (mg AA/100 mL)	TPC (mg GAE/L)	AOC (% INHIBITION)	NEBI	HMF (mg/L)
Untreated	11.7	2.72	7.0	1266	252.0	92.4	0.3	0.5
Thermal	11.8	2.79	1.1	~1000	133.6	89.4	1.7	8.0
PEF	11.7	2.76	1.0	~1200	172.6	~92.4	0.9	1.5

AA, ascorbic acid; AOC, antioxidant capacity; CFU, colony-forming unit; GAE, gallic acid equivalent (standard polyphenol); HMF, 5-hydroxymethyl-2-furfural (undesirable organic byproduct formed by dehydration of certain sugars mainly due to heat); NEBI, nonenzymatic browning index (derived color parameter); TPC, total phenolic content; TSS, total soluble solids.

maintained for 60 s. The juice was immediately cooled after treatment and stored at 4°C until further analysis.

Observations: The effect of processing condition on the count of *Z. bacilii* and other quality parameters of amla juice has been summarized in **Table 3.1**. It can be observed that both PEF and thermal processing did not affect total soluble solids (TSS) and the pH of the amla juice. A reduction of 6 log cycles in *Z. bacilii* was detected after PEF treatment, which was comparable to thermal inactivation. However, PEF proved to be better in retaining vitamin C, total phenolic content (TPC), and antioxidant capacity (AOC). PEF treatment showed less browning in color and less undesirable 5-hydroxymethyl-2-furfural (HMF) than thermal treatment. Overall, PEF proved to be a better processing alternative than conventional thermal pasteurization of the juice.

3.5 Challenges

PEF has shown encouraging results as a non-thermal processing technology that can replace conventional thermal pasteurization in the future. PEF treatment can achieve adequate microbial inactivation and enzyme inactivation (for some instances) with much better retention of nutrients, bioactives, and organoleptic properties (like appearance and taste) than thermal treatment. Consumers must also be made aware of the novel non-thermal treated products considered minimally processed products. They provide naturally fresh products that will promote their value in the market. However, enzyme inactivation still requires further optimization, support of other non-thermal or mild thermal technology, and more research. There can be a possibility of electrochemical reaction occurring at the point of contact between the electrode and the food material, which leads to a decrease in electrode life and create toxicity problem in the food sample. The commonly used steel electrode may face the problem of electrochemical reactions. Using other types of electrodes like carbon electrodes or using a modified pulse generator can help avoid or minimize the initiation of electrochemical reactions. The initial cost of equipment for PEF is very high but at the same time incurs a lower running cost. It is also true that designing a continuous processing line on an industrial scale with high uniformity in processing comparable to a batch process is quite challenging.

3.6 Summary

- In PEF processing, the liquid or semisolid food product is passed between two electrodes that allow the material to experience high-energy electric fields for short duration pulses (nanoseconds to microseconds).
- PEF is a non-thermal food processing technology that inactivates microorganisms using intense short electric pulses with high retention of nutrients and sensory properties.
- Dielectric rupture and electroporation are the two major theories proposed as the mechanisms for microbial inactivation by PEF.
- To achieve dielectric rupture of the microbial cell membrane, the electric field having voltage more than or equal to the threshold TMP is required.
- Electroporation can be achieved by the applied electric field having voltage more than the gating potential of the natural pores, pumps, and protein channels of the membrane.
- To do irreversible damage to the microbial cell and overcome its natural viscoelasticity ability to reseal the new or widened pores, sufficient exposure time and magnitude of PEF is crucial.
- Localized momentary joule heating and electro-conformational changes may inactivate enzymes.
- The effect of PEF on food depends on their respective ion concentration, pH, capacitance, and dielectric properties.

- Parallel type electrode set-up are majorly preferred because of their convenient design and uniform electric field exposure for the sample product.
- The square type waveforms and the bipolar type pulses of PEF are considered to be most the efficient.

3.7 Solved Numerical

1) A PEF treatment chamber is filled with liquid food material. The parallel plate stainless steel electrodes had a length of 162 mm and 5.2 mm width and were separated by a gap of 2 mm. Food material was given a treatment of 100 pulses in exponential decay waveform and kept under a voltage of 8 kV, maintaining an energy density distribution of 0.499 MJ·m^{-3} in the chamber. Find out the product's dielectric constant or the relative permittivity (ε_r). Consider the two given equations of energy density (Q, J·m^{-3}) and capacitance (C, Farad (F)).

$$Q = N\frac{C \cdot V^2}{2A \cdot d} \qquad C = \varepsilon_r \varepsilon_o \frac{A}{d}$$

Where, N = number of pulses, V = voltage, A = surface area of the electrode, d = distance between the two electrodes, and ε_o = absolute permittivity of free space = 8.854×10^{-12} F·m^{-1}.

Solution: Given information:-

Energy density, $Q = 0.499$ MJ/m^3 = 4.99×10^5 J·m^{-3}
Voltage, $V = 8$ kV = 8000 V
Number of pulses, $N = 100$
Length of an electrode, $L_e = 162$ mm = 162×10^{-3} m
Breadth of an electrode, $b_e = 5.2$ mm = 5.2×10^{-3} m
Distance between two electrodes, $d = 2$ mm = 0.002 m
Calculating area of each electrode, $A = L_e \times b_e = (162 \times 10^{-3}) \times (5.2 \times 10^{-3}) = 8.42 \times 10^{-4}$ m^2
Calculating capacitance (C) using the equation:

$$Q = N\frac{C \cdot V^2}{2 \cdot A \cdot d} \quad => 4.99 \times 10^5 = 100 \times \frac{C \times (8000)^2}{2 \times (8.42 \times 10^{-4}) \times 0.002}$$

$$=> C = 2.63 \times 10^{-10} \text{ F}$$

Finally, calculating the dielectric constant (ε_r) of the unknown liquid food material using the equation:

$$C = \varepsilon_r \cdot \varepsilon_o \frac{A}{d} \implies 2.63 \times 10^{-10} = \varepsilon_r \times (8.854 \times 10^{-12}) \frac{8.42 \times 10^{-4}}{0.002}$$

$$\implies \varepsilon_r = 70.56$$

2) A juice inoculated with *E. coli* was treated using PEF technology. The PEF chamber was filled with 4 mL of the juice, having an initial microbial count of 10^6 (i.e., 6 log cycles). The treatment conditions were; pulse duration = 36 μs, electric field = 20 kV/cm, treatment time = 36–1080 μs. It was observed that the inactivation of *E. coli* followed Weibull distribution kinetics. The data of the survival fraction has been given below for three conditions.

TREATMENT TIME (μs)	SURVIVAL FRACTION
36	0.220
400	0.054
1080	0.002

Calculate the average shape parameter (β) and scale parameter (δ).

Solution: Shape parameter (β) and scale parameter (δ) need to be calculated from the given data of treatment time (t) and survival fraction (s) and using the Weibull model equation:

$$-\ln(s) = \left(\frac{t}{\delta}\right)^{\beta} \implies \ln(-\ln(s)) = \beta \cdot \ln\left(\frac{t}{\delta}\right)$$

Equation A: $\ln(-\ln(0.22)) = \beta \cdot \ln\left(\frac{36}{\delta}\right)$

$$\implies 0.415 = \beta \cdot \{\ln(36) - \ln(\delta)\}$$

Equation B: $\ln(-\ln(0.054)) = \beta \cdot \ln\left(\frac{400}{\delta}\right)$

$$\implies 1.071 = \beta \cdot \{\ln(400) - \ln(\delta)\}$$

Equation C: $\ln(-\ln(0.002)) = \beta \cdot \ln\left(\frac{1080}{\delta}\right)$

$$\implies 1.827 = \beta \cdot \{\ln(1080) - \ln(\delta)\}$$

Operation B – A:

$$1.071 = \beta \cdot \{\ln(400) - \cancel{\ln(\delta)}\}$$
$$-[0.415 = \beta \cdot \{\ln(36) - \cancel{\ln(\delta)}\}]$$
$$\overline{0.656 = \beta \cdot \{\ln(400) - \ln(36)\}}$$

$$\Rightarrow \beta = 0.272$$

Operation C – B:

$$1.827 = \beta \cdot \{\ln(1080) - \cancel{\ln(\delta)}\}$$
$$-[1.071 = \beta \cdot \{\ln(400) - \cancel{\ln(\delta)}\}]$$
$$\overline{0.756 = \beta \cdot \{\ln(1080) - \ln(400)\}}$$

$$\Rightarrow \beta = 0.761$$

Operation C – A:

$$1.827 = \beta \cdot \{\ln(1080) - \cancel{\ln(\delta)}\}$$
$$- [0.415 = \beta \cdot \{\ln(36) - \cancel{\ln(\delta)}\}]$$
$$\overline{1.412 = \beta \cdot \{\ln(1080) - \ln(36)\}}$$

$$\Rightarrow \beta = 0.415$$

Average shape parameter, $\beta = \frac{0.272 + 0.761 + 0.415}{3} = 0.483$

Now calculating δ.
From Equation A: $0.415 = 0.483 \cdot \{\ln(36) - \ln(\delta)\} \Rightarrow \delta = 15.25 \ \mu s$
From Equation B: $1.071 = 0.483 \cdot \{\ln(400) - \ln(\delta)\} \Rightarrow \delta = 43.55 \ \mu s$
From Equation C: $1.827 = 0.483 \cdot \{\ln(1080) - \ln(\delta)\} \Rightarrow \delta = 24.58 \ \mu s$

Average scale parameter, $\delta = \frac{15.25 + 43.55 + 24.58}{3} = 27.86 \ \mu s$

3) Effect of different PEF processing conditions on ascorbic acid (AA) content of a fruit juice-milk beverage (milk-shake) was observed. A constant electric field strength (E) of 25 kV·cm^{-1} having square type waveform of pulses was applied, which possess a constant pulse duration (τ; ON-time of each wave when energy is delivered to the sample) of 2.5 μs per pulse. It has been observed that at different treatment times (t, μs), the final temperature (T_f, °C) of the sample increased, which affected the conductivity (σ, mS·cm^{-1}; here 'S' stands for the

unit Seimens $= \Omega^{-1}$) of the beverage sample. Calculate the total energy $(Q_E, J \cdot mL^{-1})$ received by the beverage sample for each condition and the mean degradation rate constant $(k, \mu s^{-1})$ of AA, assuming first-order kinetics. The details of the experiment have been given below.

S.NO.	$t\,(\mu s)$	$T_f(°C)$	$\sigma\,(mS \cdot cm^{-1})$	AA $(mg \cdot 100\ mL^{-1})$
1	0	22.0	2.99	22.72
2	130	44.5	4.62	21.22
3	310	59.0	5.67	20.92

Solution: Given information are; $E = 25 \times 10^3$ V·cm^{-1} and $\tau = 2.5\ \mu s$ per pulse

Total energy delivered to the sample,

$$Q_E = n \cdot E^2 \cdot \sigma \cdot \tau = \frac{t}{\tau} \cdot E^2 \cdot \sigma \cdot \tau$$

[Note the unit balance: $(V \cdot cm^{-1})^2 \cdot (S \cdot cm^{-1}) \cdot (s) = (V^2 \cdot \Omega^{-1}) \cdot s \cdot cm^{-3} = (J \cdot s^{-1}) \cdot s \cdot mL^{-1} = J \cdot mL^{-1}$]

Here, n is the number of pulses;

$$n = \frac{t}{\tau} \quad (OR) \quad n = t\nu$$

[Note: ν = frequency (Hz)]

a. At $t = 0\ \mu s$; $Q_E 1 = 0\ J \cdot mL^{-1}$
b. At $t = 130\ \mu s$; $Q_E 2 = (130/2.5) \times (25 \times 10^3)^2 \times (4.62 \times 10^{-3}) \times (2.5 \times 10^{-6}) = 375.38\ J \cdot mL^{-1}$
c. At $t = 310\ \mu s$; $Q_E 3 = (310/2.5) \times (25 \times 10^3)^2 \times (5.67 \times 10^{-3}) \times (2.5 \times 10^{-6}) = 1098.56\ J \cdot mL^{-1}$

Calculating degradation rate constant of AA due to PEF treatment, using the first order kinetic model; $\ln(C/C_0) = -k \cdot t$

Here, C_0 is the AA concentration at $t = 0$ and C is the AA concentration at $t > 0$

a. At $t = 130\ \mu s$;

$$\ln\left(\frac{21.22}{22.72}\right) = -k_1(130) \implies k_1 = 5.25 \times 10^{-4}\ \mu s^{-1}$$

b. At $t = 310\ \mu s$;

$$\ln\left(\frac{20.92}{22.72}\right) = -k_2(310) \implies k_2 = 2.66 \times 10^{-4}\ \mu s^{-1}$$

Average k for $AA = (k_1 + k_2)/2$

$$=> k = ((5.25 \times 10^{-4}) + (2.66 \times 10^{-4}))/2$$
$$=> k = 3.96 \times 10^{-4} \mu s^{-1}$$

[Note: The rate constant can also be calculated from the slope of the semi-log plot between time and AA concentration.]

3.8 Multiple Choice Questions

1. What are the possible reasons for microbial inactivation in food during PEF treatment.
   a. Membrane rupture by electro-compression
   b. Electro-conformational changes to the lipid bilayer of the cell membrane
   c. Formation of hydrophilic pores and enlargement of the existing ones
   d. All of the above

2. Assuming a spherical shaped microbial cell of 0.5 μm diameter, what is the minimum electric field (E) required to overcome the threshold TMP of about 1.2 V (in the direction of applied E) and the reversible effects of cell's natural viscoelastic restoration property?
   a. 26.7 kV/cm
   b. 32.0 kV/cm
   c. 64.0 kV/cm
   d. 53.3 kV/cm

3. Which of the following product qualities will enhance the microbial inactivation by PEF treatment? Product qualities are (1) High concentration of ions, (2) High electrical resistivity, (3) High electrical conductivity, (4) High dielectric constant, and (5) Low pH.
   a. 1, 3, & 5
   b. 1, 2 & 3
   c. 2, 3 & 4
   d. 2, 4, & 5

4. Which type of pulse waveform and electrode configuration is considered to be most efficient and suitable for continuous industrial PEF processing, respectively?
   a. Square waveform & parallel plate electrodes
   b. Exponential decay waveform & coaxial cylindrical electrodes
   c. Square waveform & coaxial cylindrical electrodes
   d. Exponential decay waveform & parallel plate electrodes

5. Which one of the following is not relevant to PEF assembly and food processing?
   a. Heat of compression
   b. High voltage electrodes
   c. Pulse generator
   d. Electroporation

6. The gap between successive discharge or consecutive pulses during which no electrical energy is released into the product can vary from _____.
   a. One second to minutes
   b. One millisecond to seconds
   c. One microsecond to milliseconds
   d. One nanosecond to microseconds

7. How the electric field strength (E) is related to voltage supply (V) and the distance (d) between the two electrodes?
   a. V is directly proportional, and d is inversely proportional to E
   b. V is inversely proportional, and d is directly proportional to E
   c. Both V and d are directly proportional to E
   d. Both V and d are inversely proportional to E

8. What is the approximate range of dielectric constant (ε) in the case of food materials?
   a. 20–40
   b. 40–60
   c. 60–80
   d. 80–100

9. In microorganisms _____ is approximately the natural TMP across the membrane and _____ is approximately the reported threshold TMP required for the loss of membrane integrity and localized membrane breakdown
   a. 50 mV, 1 V
   b. 10 mV, 1.5 V
   c. 50 mV, 1.5 V
   d. 10 mV, 1 V

10. The yeast cells are very susceptible to PEF treatment because of the cells possess:
   a. Good conductivity and poor capacitance
   b. Good conductivity and good capacitance
   c. Poor conductivity and good capacitance
   d. Poor conductivity and poor capacitance

3.9 Short Answer Type Questions

a. How is the cell permeability of the microbial cell membrane affected by PEF treatment?
b. Define TMP and its role in microbial inactivation by PEF.
c. Define the strength or intensity of a constant electric field (E). How is it related to the TMP for inactivating a spherical microbial cell?
d. How may PEF processing affect enzymes in food?
e. What must be done to overcome the cell membrane's resealing or healing ability?

3.10 Descriptive Questions

a. Discuss the theories regarding PEF based microbial inactivation.
b. How is the mechanism of microbial inactivation by PEF affected by capacitance of food, the direction of the electric field, microbial cell diameter, and cell membrane's natural viscoelasticity?
c. Does the state or type of food product matter during PEF treatment? What is the impact of PEF on nutritional and sensory aspects of food?

d. With a schematic, describe the effect of equipment type and mode of operation on the efficacy of PEF processing.

e. Point out the challenges associated with PEF processing of food materials.

3.11 Numerical Problems

1) Changes in total carotenoids (TC) concentration during cold storage (4°C) of an orange juice-milk beverage treated with PEF was studied for 42 days. Initially, TC of the beverage sample was 3478.8 $\mu g \cdot L^{-1}$, which got reduced to 2814.1 $\mu g \cdot L^{-1}$ after 21 days, and on the last day, the TC was noted as 2320.2 $\mu g \cdot L^{-1}$. Based on the three data points, develop the first- and second-order kinetic models and comment on the best fit model as per minimum total error.

2) In many cases, ascorbic acid (AA) content has been taken as the quality marker for treated juices. PEF processing of grapefruit juice was explored at an electric field strength (E) of 35 kV $\cdot$ cm^{-1} with 0.4 × 10^6 Hz frequency of pulses for two different treatment times (t). Due to different levels of temperature rise of the treated juice, conductivity (σ) of the grapefruit juice varied. The details of the analysis have been described below.

S.NO.	t (μs)	σ (mS $\cdot$ cm^{-1})	AA (mg $\cdot$ L^{-1})
1	0	1.23	362.31
2	80	3.04	344.19
3	180	4.53	333.87

A modified zeroth order kinetic model was applied to obtain the degradation rate constant (k, L $\cdot$ J^{-1}). As Q_E (J $\cdot$ L^{-1}) is dependent on t, it has been put in place of t; $C = C_0 - k \cdot Q_E$. Estimate the mean rate constant for AA degradation by PEF.

References

Bansal, V., Sharma, A., Ghanshyam, C., Singla, M. L., & Kim, K. H. (2015). Influence of pulsed electric field and heat treatment on *Emblica officinalis* juice inoculated with *Zygosaccharomyces bailii*. *Food and Bioproducts Processing*, *95*, 146–154.

Suggested Readings

Aguiló-Aguayo, I., Elez-Martínez, P., Soliva-Fortuny, R., & Martín-Belloso, O. (2012). High-intensity Pulsed Electric Field Applications in Fruit Processing. In S. Rodrigues, & F. A. N. Fernandes (Eds.), *Advances in Food Processing Technologies*. CRC Press, Taylor and Francis Group, Boca Raton, Florida, USA. pp. 149–185.

Huang, K., & Wang, J. (2009). Designs of pulsed electric fields treatment chambers for liquid foods pasteurization process: A review. *Journal of Food Engineering*, *95*(2), 227–239.

Syed, Q. A., Ishaq, A., Rahman, U. U., Aslam, S., & Shukat, R. (2017). Pulsed electric field technology in food preservation: A review. *Journal of Nutritional Health & Food Engineering*, *6*(5), 168–172.

Zhang, H. Q., Barbosa-Cánovas, G. V., Balasubramaniam, V. B., Dunne, C. P., Farkas, D. F., & Yuan, J. T. (Eds.). (2011). *Non-thermal Processing Technologies for Food*. Wiley-Blackwell, Chichester, West Sussex, UK.

Answers for MCQs (sec. 3.8)

1	2	3	4	5	6	7	8	9	10
d	b	d	c	a	b	a	c	d	c

The answer of *Q2* is option 'b' according to Eq. 3.2

4

ULTRAVIOLET AND PULSED LIGHT TREATMENT

4.1 Principle and Operation

A mercury-based ultraviolet (UV) lamp generates light spectra in the range of 100–400 nm. The UV range can further be broken down into four subdivisions: UV-A (315–400 nm), UV-B (280–315 nm), UV-C (200–280 nm), and vacuum-UV (100–200 nm) (**Fig. 4.1**). There exist different types of lamps and configurations, among them mercury lamps, which are most popular and easy to handle. In a lamp filled with mercury vapors at a specific pressure, the electrons in the mercury atoms get excited when the current is allowed to pass through it and emits UV radiation. In contrast, the excited electrons return to their initial lower energy level. The photons of the UV light have enough energy to initiate various photochemical effects that involve breaking and forming chemical bonds. Still, they won't ionize the treated product, which is the same in the case of pulsed light (PL) as well. The energy of a photon can be defined using the Plank-Einstein relation (Eq. 4.1).

$$E_p = h \cdot v$$

$$v = \frac{c}{\lambda}$$

(4.1)

Where,
E_p = Energy of photon (kJ/Einstein or electronvolt (eV))
h = Planks constant = 4.14×10^{-15} eV·s = 6.63×10^{-34} J·s
v = Frequency (Hz)
c = Speed of light = 3.0×10^8 m/s
λ = Wavelength (m)

Eq. 4.1 indicates that photons of light having a shorter wavelength and higher frequency will carry higher energy, and higher will be its penetration. On the other hand, PL is a multi-spectrum radiation

DOI: 10.1201/9781003199809-4

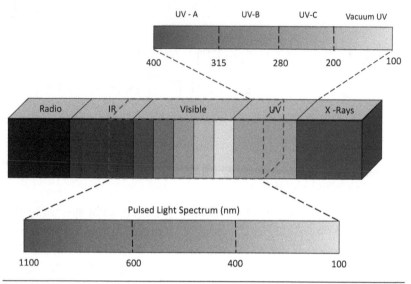

Figure 4.1   Light spectrum distribution of UV and PL.

(100–1100 nm), which includes UV (200–380 nm), visible range (380–700 nm), and near-infrared (near-IR, 700–1100) range. The PL flash lamp is filled with an inert gas, such as argon, krypton, or xenon. However, xenon is preferred more because of its better output. Under the influence of passing current, the pressurized xenon gas is excited to release a polychromatic light. The radiation is delivered in a pulsating form by supporting the pulse forming network attached to the circuit through which energy is supplied to the lamp. The significant germicidal property of the PL primarily comes from the 21–40% UV fraction.

4.2 Mechanism of Quality Changes in Food

4.2.1 Microbial Inactivation

The primary mechanism associated with microbial inactivation by both UV and PL is the photochemical effect of radiation on microorganisms' DNA and RNA. The photons of the UV range have high energies absorbed by the microorganisms that lead to the formation of mainly pyrimidine dimers (dimers of cytosine or thymine) in the single or double-stranded DNA structures (**Fig. 4.2**). Altering the DNA structure makes the cell unable to replicate, and therefore kills the microorganism clonogenically. A microorganism that has lost its

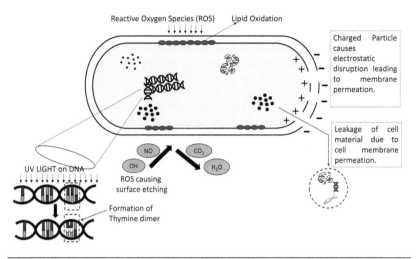

Figure 4.2   Microbial inactivation mechanism due to PL or UV processing.

replication ability cannot multiply and cause infection to any human body. However, some microorganisms can repair a fraction of their damaged DNA with the help of enzymes like DNA photolyase under daylight exposure (310–700 nm); this phenomenon is known as photo-reactivation. This can be prevented by exposing the food to a higher dose of UV or PL to incur irreversible damage to the DNA and storing the treated product in dark conditions. Apart from the photo-chemical effect, PL irradiation involves other phenomena responsible for microbial inactivation, including photothermal and photothermal effects. In the case of photothermal effects, the presence of IR range in the PL contributes toward the significant temperature rise of the product, which travels from the outside irradiated surface to the other parts of the product; this causes vaporization of water, cell shrinkage, and membrane damage.

More prolonged PL irradiation can increase product temperature significantly, which is not desirable while maintaining non-thermal conditions. On the other hand, the photophysical effect, also known as the third type of effect, might have been initiated by photochemi-cal and photothermal effects. It involves damage to the cell membrane or wall, shrinking the cytoplasmic membrane, rupturing mesosomes, and leakage of cytoplasmic materials. Overall, PL-based inactivation is a multi-target process in which all the phenomena may together have a synergistic impact.

4.2.2 Effect on Spoilage Enzyme and Nutritional Quality

Inactivation of enzymes occurs mainly due to photochemical effects of PL or UV light. Upon exposure to the UV fraction, the amino acids (cysteine, methionine, phenylalanine, tryptophan, and tyrosine) undergo electrical excitation to produce thiol groups (sulfanyl group 'R-SH'), which further cause breakage of disulfide (part of the tertiary structure) and peptide bonds and eventually damage the secondary protein structures of the enzymes (α helix and β sheet). Photo-oxidation of amino acid residues and formation of unstable thiol groups may react to build covalent cross-linkages that lead to aggregation or promote backbone cleavage of the secondary protein structure via fragmentation. Even a slight alteration at the enzyme's active site can render it inactive. Apart from this, the loss of nutrients and bioactive and organoleptic properties was insignificant for a few cases. However, researchers who have observed any loss have stated the reason to be the photo-oxidation phenomenon. Interestingly, it has also been reported that in some cases, the total phenolic content (TPC) and the antioxidant capacity increased after treatment, probably due to depolymerization and improved extractability. So, depending on the processing conditions, food product, types, and concentration of various polyphenols and antioxidants, the phenolic content and antioxidants may increase, decrease, or remain unchanged. Overall, a more detailed investigation of the associated mechanisms for enzyme inactivation and changes in nutritional qualities is still necessary.

4.3 Equipment Functioning

A schematic of an annular type of UV reactor has been shown in **Fig. 4.3**. UV radiation produced by low and medium-pressure mercury arc lamps contains a mixture of mercury and argon (inert gas) vapors held inside a UV transmitting quartz or silica tube. The tube ends are fitted with two electrodes, generally made of tungsten metal mixed with other alkaline earth metals (helps in arc formation). The low-pressure mercury arc lamp produces monochromatic light (254 nm), and the medium pressure mercury arc lamp generates polychromatic light in UV and visible range.

Fig. 4.4 shows the schematic of a laboratory model batch assembly of PL treatment. The major components of PL assembly are controller

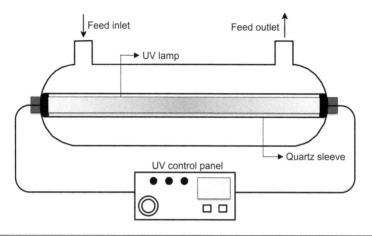

Figure 4.3 Continuous ultraviolet reactor with an annular design.

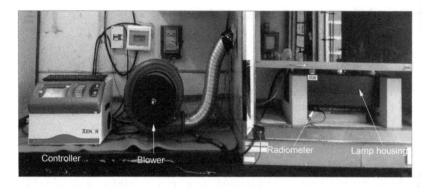

Figure 4.4 A laboratory scale batch type pulse light processing system installed at Food Engineering and Technology Department, Institute of Chemical Technology, Mumbai, India.

unit, lamp housing, blower, sample holder, and radiometer. The PL system is equipped with a high-voltage power supply that temporarily stores electrical energy in the capacitors. The capacitors help to power up the lamp with its stored energy. The release of light in pulses of specific shapes is facilitated by a pulse forming network of capacitors and inductors. A high-voltage trigger delivers the pulsating energy to the flash lamp. The lamp is enclosed inside a metallic case (lamp housing) having a quartz window. The lamp housing is supplied with an air blower to keep the lamp safe from overheating. Both PL and UV require a separate optical sensor (radiometer) to measure the irradiated product's fluence or energy per unit area. The schematic of a continuous PL processing system has been shown in **Fig. 4.5**.

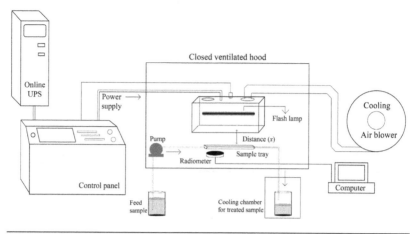

Figure 4.5   Continuous-type pulse light processing system.

4.4 Critical Processing Factors

4.4.1 Type of Product

Even though PL has greater penetration than UV, both PL and UV radiations have their limitations in penetration. Optically opaque and translucent material significantly hinders the passage of germicidal rays because light experiences refraction, reflection, scattering, and absorption as it falls onto the product. The product needs to have low absorbance (with fewer colored components and low turbidity) and good transmittance to obtain adequate microbial or enzyme inactivation. Clear liquid products have a lower concentration or smaller size of particulate matter, thus making them suitable for UV or PL treatment. Examples are drinking water, coconut water, and clarified apple juice. Particles or turbidity can shield the microorganisms and spoilage enzymes from inactivating. While irradiating, the product's thickness or the liquid film's thickness is also critical. The lower the thickness, the better the penetration up to the bottom of the product.

According to the Beer-Lambert law, the absorbance (A) depends on the concentration of specific radiation absorbing species (c, mol/L), extinction coefficient or molar absorptivity (ε, L/mol·cm), and path length from the light source to the products bottom surface or one end to another (d, cm) (Eq. 4.2).

$$A = \varepsilon \cdot c \cdot d \qquad (4.2)$$

Like the extinction coefficient, the absorption coefficient (α, cm^{-1}) defines the product or medium's intrinsic property regarding absorption characteristics. α having a base e (α_e) (Eq. 4.3) or base 10 (α_{10}) is known as Nefarian and logarithmic absorption coefficient, respectively.

$$\alpha_e = 2.303 \frac{A}{d}$$
(4.3)

$$\alpha_e = \frac{1}{\lambda_d}$$
(4.4)

$$E_x = E_o \cdot \exp(-\alpha_e \cdot x)$$
(4.5)

Where λ_d is the penetration depth, E_x is the fluence rate (intensity of light or flux rate of light energy) at x distance from the light source, and E_o is fluence rate $x = 0$. The penetration depth of the light represents the reachable depth in the medium of the sample or 'sample + air' at which the initial intensity (fluence rate) falls by 95–99%. Apart from these, a product's pH and acidity may also affect the efficacy of the treatment.

4.4.2 Relative Positioning of the Product

According to the equations (4.3 to 4.5), the distance or relative positioning of the product as per the treatment chamber and the light source is also an essential factor. The product should not be kept far from the light source for the more substantial impact of UV or PL irradiation.

4.4.3 Target Microorganism or Enzyme

Microorganisms and enzymes may have variable resistance toward UV and PL treatment. The susceptibility of Gram-negative bacteria (such as *Escherichia coli* and *Pseudomonas* spp.) is the highest toward UV or PL, Gram-positive bacteria (such as *Listeria* and *Bacillus* spp.) have moderate susceptibility, and fungal spores (such as *Saccharomyces cerevisiae*) are the most resistant. However, in some cases, a different order of susceptibility has also been observed. For example, *S. cerevisiae* was most sensitive, and *Listeria monocytogenes* were the least sensitive toward PL. The additional effect of photothermal effect (other than photochemical effect) of PL can be quite potent against fungus, such as *Aspergillus niger*. Fungal or bacterial spores are highly resistant to both UV and PL. So, different microorganisms in various buffer solutions or a relevant food matrix can show variable resistance. On the other hand, enzymes may show different susceptibility depending on the source and type. The initial microbial load or concentration of enzymes in the product also influences the inactivation process significantly. A higher initial count of viable cells or higher initial concentration of enzymes may cause a shadowing effect that will reduce the uniformity of the irradiation process and finally cause the inactivation by UV or PL to reduce. Microorganisms show different rates of inactivation under UV or PL, which can be quantified using kinetic equations, such as first-order kinetics (Eq. 4.6).

$$s = \frac{N}{N_o} = \exp(-k \cdot E_x \cdot t) \qquad (4.6)$$

Where s is the survival fraction, N_o is the initial microbial count, N is the count at time t, and k is the inactivation rate constant. First-order kinetics is the simplest form of a kinetic equation.

4.4.4 Equipment and Mode of Operation

To utilize the equipment or processing technology, controlling its parameters is essential. The basic parameters for UV or PL treatment are as follows:

- Fluence (F_o): It is the total light energy delivered per unit area or light energy flux ($J \cdot m^{-2}$) (Eq. 4.8).
- Fluence rate (E_o): The light energy per unit area and per unit time ($J \cdot m^{-2} \cdot s^{-1}$).
- Treatment time (t): Exposure time (s).
- Pulse period or pulse width (t_p): During the ON-OFF time of PL irradiation, the timespan of ON time (μs) when energy is delivered is called the pulse width or period.
- Frequency (ν): It is the pulse repetition rate that can be seen as the number of pulses per unit time (Hz, 'the number of pulses $\cdot s^{-1}$'), applicable for PL (Eq. 4.7).
- Peak power (P_o): It is the total radiant power ($J \cdot s^{-1}$ or W, i.e., Watt), and for PL, the peak power can be measured as the

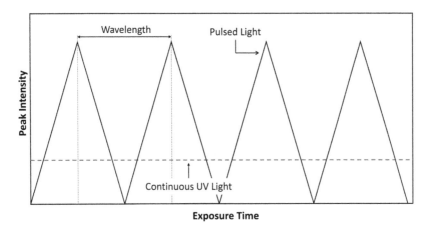

Figure 4.6   Comparison between PL and UV in terms of peak power.

energy of the peak pulse per unit pulse duration or width of that pulse. Power delivered by the lamp can also be expressed as the product of fluence rate and surface area (S_A, m²) of the lamp (Eq. 4.9). While comparing, it can be observed that PL delivers light energy in pulses, whereas UV delivers light energy constantly and continuously (**Fig. 4.6**). Therefore, in the case of UV, power can simply be estimated as light energy delivered per unit time.

$$v = \frac{1}{t_p} \tag{4.7}$$

$$F_o = E_o \cdot t = (\text{fluence per pulse}) \times (\text{number of pulses}) \tag{4.8}$$

$$P_o = E_o \cdot S_A \tag{4.9}$$

Changing the voltage supply to the light source can modify F_o, E_o, and P_o of the radiation. In addition, the voltage supply for PL's lamp may also cause some alterations in the proportion of the constituent (UV, visible, and IR) of its multi-spectrum light. Higher the voltage, more will be the delivered F_o and E_o, which would increase microbial or enzyme inactivation to some extent depending on the target microorganism or enzyme. Similarly, increasing exposure time will cause the product to receive more fluence (F_o) and cause more inactivation of microorganisms and enzymes up to a certain level. Besides, increasing the frequency can deliver equivalent fluence in a shorter time than a lower frequency setting. In the case of a continuous flow treatment, controlling the flow to obtain adequate exposure time should also be taken care of.

4.5 Practical Example

We will consider one of the UV- and PL-based research work as a practical example for better understanding. In research conducted by Caminiti et al. (2012), a juice blend made of orange and carrot juices (1:1 v/v) was treated with non-thermal and thermal processing conditions. Among the non-thermal techniques, two were UV and PL treatment. The authors studied the effect of processing on pH, total soluble solids (TSS, °Brix), total color change (ΔE), nonenzymatic browning index (*NEBI*), TPC, and pectin methylesterase (PME) activity. PME is an undesirable spoilage enzyme that causes phase separation in fruit juices.

UV treatment: The juice blend was irradiated with UV for 1 min, delivering an overall dose (total fluence, F_0) of 10.62 J/cm^2.

PL treatment: A total dose of 3.3 J/cm^2 was delivered to the juice blend in which each pulse carried the energy of 1.21 J/cm^2. The other settings were 3 Hz pulse frequency and 20.8 mL/min flow rate.

Thermal pasteurization: The juice blend was pasteurized at 72°C for 26 s in a tubular heat exchanger. The heat exchanger took about 3.5 min (come up time) to increase the temperature of the product up to 72°C.

Observations: The results mentioned in **Table 4.1** show that no significant change in pH and TSS was visible after various processing conditions. Again, no change in the NEBI value was observed

Table 4.1 Effect of Various Pasteurization Treatments on PME Enzyme Activity and Various Physicochemical Attributes in the Juice

TREATMENT	pH	TSS (°Brix)	ΔE	NEBI	TPC (mg GAE/L)	PME (% RESIDUAL ACTIVITY)
Untreated	3.75	9	0	0.197	505	100
Thermal	~3.75	~9	2.91	0.197	500	8
PL	~3.75	~9	0.49	0.206	501	113
UV	~3.75	~9	0.74	0.202	498	82

GAE, gallic acid equivalent (standard polyphenol); *NEBI*, nonenzymatic browning (derived color parameter); PME, pectin methylesterase; TPC, total phenolic content; TSS, total soluble solids; ΔE, total color change.

after thermal pasteurization compared to the untreated juice blend, but a slight change in PL and UV treated was seen. A significant change in the ΔE was discovered in the case of thermal pasteurization compared to the UV and PL treated products. This shows that thermal treatment can destroy color pigments in the juice blend at a higher level than non-thermal techniques, such as UV and PL. NEBI can also increase ΔE, but NEBI was not significantly affected in this case. Therefore, the increase in ΔE can be attributed to the heat-induced destruction of color pigments. TPC was also not much affected by all the processing techniques, but UV treatment reduced the TPC slightly more than other techniques. A 92% reduction in the PME activity was obtained from thermal processing, which is the highest. At the same time, UV treatment caused 18% inactivation in PME. In the case of PL, the PME activity increased by 13%, which was unexpected and undesirable. Other researchers have also reported similar results for low-intensity conditions of UV treatment. However, some enzyme inactivation has been reported after PL treatments. Overall, single UV and PL-based processing cannot inactivate enzymes, and thermal pasteurization harms the beverage's color.

4.6 Challenges

PL and UV are both non-thermal techniques; they are environment friendly and have great potential to replace thermal pasteurization methods. While handling PL-based processing, care needs to be taken to keep the exposure time short to avoid temperature rise; if a longer treatment time is necessary, then the initial temperature of the product should be kept as low as possible, or some online-cooling mechanism for the product must be arranged. Even though PL has a higher penetration depth than UV light, both radiations have limited penetration depending on the sample color, transmittance, viscosity, turbidity, and particulate matter. Therefore, liquids with poor transmittance are unsuitable for UV or PL treatment. Semisolid and solid food products can only be surface disinfected using UV or PL irradiation. This can be overcome by making the opaque liquid or semisolid products flow in thin film or by following combination (hurdle technologies) of other non-thermal or mild thermal techniques. The majority of the pathogens can be easily killed or inactivated by UV or PL exposure. Still, few spoilage enzymes, resistant microorganisms like bacterial or fungal spores or spore formers may be complex, but longer treatment time or hurdle techniques can make inactivation easier. The PL system needs extra machinery and complex electrical circuit support, it is costlier than the UV systems. However, the running cost for both UV and PL might be lower than thermal processing.

4.7 Summary

- UV and PL are radiation-based non-thermal technologies in which UV is continuous, and PL is pulsating in nature.
- UV lamps contain mercury vapors, and PL lamps majorly contain xenon gas.
- UV is monochromatic or polychromatic over a shorter range of spectrum (100–400 nm), and PL is broad-spectrum polychromatic light (100–1100 nm) because the spectrum of PL consists of UV, visible, and near-infrared range.
- UV-C light, specifically 254 nm, is the most germicidal radiation in both UV and PL. Treated sample must be stored in the dark to avoid possible photoreactivation.
- UV inactivates microorganisms and enzymes by photochemical effect, and PL achieves them by both photochemical and photothermal effects.
- PL has higher penetration and shows higher lethality than UV, but a more extended treatment period of PL can result in a temperature rise of the food sample.
- The total light energy delivered to per unit area of the food sample is quantified as fluence or dose ($J \cdot m^{-2}$), and the fluence delivered per unit time is expressed as fluence rate ($J \cdot m^{-2} \cdot s^{-1}$).
- Liquid food samples with a high concentration of particles and poor transmittance are not suitable for UV or PL treatment.
- Only surface disinfection is possible for opaque liquids, semisolid, and solid food products.
- UV systems have a more convenient design, continuous, and batch processing methods and are cheaper compared to PL systems.

4.8 Solved Numerical

1) Orange juice is treated in an annular UV-C reactor to reduce the aerobic mesophilic count (AMC). The annular (ring-shaped) volume available for the juice flow was 0.675 L, and the juice flowed at 4000 L/h in a thin film with negligible distance from the central lamp. The UV lamp had an effective outer surface area of 661.93 m² that delivers light at 25.5 W supplied power to the lamp. After treatment, a reduction of 3.31 $\log_{10}$ was observed. Estimate the total fluence or dosage received by the juice and the inactivation rate constant $(cm^2 \cdot J^{-1})$ of the aerobic mesophiles as per first-order kinetics.

Solution: Given information:

Volume available for the orange juice to flow, $V_J = 0.675$ L $= 0.000675$ m³

Juice flowrate, $\dot{m} = 4000$ L/h $= 1.111$ L·s⁻¹

UV-C lamp's output power, $P_o = 25.5$ W $= 25.5$ J·s⁻¹

The effective outer surface area of the lamp, $S_A = 661.93$ m²

$\log_{10}$ of survival fraction of AMC, $\log(s) = \log(N/N_o) = -3.31$

Calculations:

Retention time (t) of the juice while undergoing treatment $= \frac{V_J}{\dot{m}}$

$= \frac{0.675 \text{ L}}{1.111 \text{ L/s}} = 0.608$ s

Fluence rate (E_o) or light intensity $= \frac{P_o}{S_A} = \frac{25.5 \text{ W}}{661.93 \text{ cm}^2} = 0.0385$ W/cm²

$= 38.5$ mW·cm⁻²

Fluence (F_o) or dosage $= E_o \times t = (38.5$ mW/cm²$) \times (0.608$ s$) = 23.408$ mJ·cm⁻²

$$\log(s) = \log\left(\frac{N}{N_o}\right) = -k \cdot E_x \cdot t \text{ \{First order kinetic equation\}}$$

Negligible distance was assumed between lamp's surface and juice layer; thus $E_x = E_o$

$$-3.31 = -k \times (0.0385 \text{ J/(s·cm}^2)) \times (0.608 \text{ s})$$

Inactivation rate constant, $k = 141.405$ cm²·J⁻¹

2) To design a PL treatment system, the flash-lamp's lifetime must be specified. A xenon flash-lamp has a quartz tubing coefficient of 21,000 J·cm⁻²·s⁻⁰·⁵. Arc length and a bore diameter of the lamp are

10 cm and 5 mm, respectively. Suppose the lamp has a lifetime of a minimum of 10^6 pulses for a width of 360 μs with a frequency of 3. Calculate the operating energy input to be maintained.

Solution: Given information:

Quartz tubing coefficient, $Q = 21000 \text{ J} \cdot \text{cm}^{-2} \cdot \text{s}^{-0.5}$
Arc length of the lamp, $l = 10$ cm
Bore diameter, $d_b = 5$ mm $= 0.5$ cm
Life $= 10^6$ pulses
Operating energy, E_o (J) $=$?
Calculations:
Pulse width, $t_p = 1/3 \times 360 \ \mu s = 120 \times 10^{-6}$ s
Single pulse explosion constant, k_e $(\text{J} \cdot \text{s}^{-0.5}) = Q \times l \times d_b = 21{,}000 \times 10 \times 0.5 = 105{,}000$
Explosion energy for lamp, E_x (J) $= k_e \cdot (t_p)^{0.5} = 105{,}000 \times (120 \times 10^{-6})^{0.5} = 1150.22$

$$\text{Life} = (E_x/E_o)^{-8.5}$$
$$\Rightarrow 10^6 = (E_o/1150.22)^{-8.5} \qquad \Rightarrow E_o = 226.41 \text{ J}$$

3) One cylindrical vessel (Φ 50 cm $\times$ 100 cm) has been half-filled with hydrogenated oil and water in 1:2 (oil: water v/v). It is planned to be exposed in front of light radiation having an energy of 10 kJ. Calculate the possible temperature rise at the bottom of a 2-mm thick layer from the top of the water layer. The reflection coefficient of water and oil are 0.65 and 0.75, respectively, and assume the rest of the transmitted energy per unit area has been completely converted to heat. Also, assume that oil and water are entirely immiscible and take extinction coefficients of water and oil as 0.1 and 0.2, respectively.

Solution: Given information:

50% of the vessel is occupied by 'oil + water' having total height, $h_t = \frac{100 \text{ cm}}{2} = 50$ cm $= 0.5$ m

Proportion of two immiscible liquids, oil and water $= 1:2$ (v/v)
Height of oil fraction, $h_o = (1/3) \times h_t = 0.5/3 = 0.167$ m
Height of water fraction, $h_w = (2/3) \times h_t = (2/3) \times 0.5 = 0.333$ m
Reflection coefficient for water, $r_w = 0.65$

Reflection coefficient for oil, $r_o = 0.75$

Extinction coefficient for water, $\alpha_w = 0.1$

Extinction coefficient for oil, $\alpha_o = 0.2$

The energy of incident radiation on 'oil + water,' $E_i = 1 \times 10^4$ J

Temperature rise at 2 mm below water layer, $\Delta T_w = ?$

Considerations: Density of water, $\rho_w = 1000$ kg/m³

Specific heat of water, $Cp_w = 4182$ J·kg⁻¹·°C⁻¹

Calculating incident fluence or dose $(F_i) = \frac{E_i}{\text{Surface area}} = \frac{1 \times 10^4}{\pi \times (0.25)^2} = 50955.4$ J·m⁻²

Oil has a lower density than water, it floats over the water layer, and the incident light approaches the oil layer first. The amount of fluence transmitted through the oil layer is F_o (J·m⁻²).

$$F_o = (1 - r_o) \cdot F_i \cdot e^{(-\alpha_o \cdot b_o)} = (1 - 0.75) \times 50955.4 \times e^{(-0.2 \times 0.167)}$$

$$= 12320.4 \text{ J·m}^{-2}$$

Now, F_o becomes the incident fluence for water layer. We need to calculate the fluence that reached 2 mm (b_{wt}) below water layer. $b_{wt} = 0.002$ m

Fluence reaching 0.002 m below water layer, $F_w = (1 - r_w) \cdot F_o \cdot e^{(-\alpha_w \cdot b_{wt})} = (1 - 0.25) \times 12320.4 \times e^{(-0.1 \times 0.002)} = 9238.4$

Volume of 0.002 m deep water, $V_w = \pi \times r^2 \times b_{wt} = 3.14 \times (0.25)^2 \times 0.002 = 0.000393$ m³

F_w is assumed to be completely converted into heat.

$$F_w = (\rho_w \times V_w) \times Cp_w \times \Delta T_w$$

$\Rightarrow 9238.4 = (1000 \times 0.000393) \times 4182 \times \Delta T_w \qquad \Rightarrow \Delta T_w = 5.63$ °C

4.9 Multiple Choice Questions

1. Select the correct combination of the listed statements (1–4) about the difference between UV and PL.

(1) The spectrum band of UV is shorter compared to PL, (2) UV lamps typically contains xenon gas, but PL lamps contain mercury vapor, (3) Light energy is delivered in a pulsating form in PL, but that is generally not the case for continuous UV,

(4) UV may have some heating effects on prolonged exposure whereas, PL doesn't show such behavior.
   a.   1 & 2
   b.   1 & 3
   c.   2 & 4
   d.   3 & 4

2. What wavelength range within the UV spectrum is most lethal for microorganisms?
   a.   315–400 nm
   b.   280–315 nm
   c.   200–280 nm
   d.   100–200 nm

3. The microbial inactivation by UV is caused by the _____ effect, and in the case of PL, it can be because of _____ effects.
   a.   photochemical, photochemical + photothermal + photophysical
   b.   photochemical, photothermal + photophysical
   c.   photothermal, photochemical + photothermal
   d.   photothermal, photochemical + photothermal + photophysical

4. Which type of microorganisms are most susceptible to UV or PL radiation?
   a.   Fungal spores
   b.   Gram-positive bacteria
   c.   Gram-negative bacteria
   d.   All types of microorganisms are equally susceptible

5. Choose the correct statement regarding the significance of the Nefarian absorption coefficient (α_e, cm^{-1}) of a medium whose relationship with the intensity of light (fluence rate) is derived from the Beer-Lambert's law.
   a.   If $\alpha_e > 1$: light intensity will deplete along its medium path. If $\alpha_e < 1$: light intensity will be enhanced along its path in the medium
   b.   If $\alpha_e > 1$: light intensity will be enhanced along its medium path. If $\alpha_e < 1$: light intensity will be depleted along its path in the medium

c. If $\alpha_e > 0$, the light intensity will be enhanced along its medium path. If $\alpha_e < 0$: light intensity will be depleted along its path in the medium

d. If $\alpha_e > 0$, the light intensity will deplete along its medium path. If $\alpha_e < 0$: light intensity will be enhanced along its path in the medium

6. The flash lamp of PL is filled with a noble gas at a specific pressure. Which gas is considered the most efficient for this purpose?
   a. Neon (Ne)
   b. Argon (Ar)
   c. Krypton (Kr)
   d. Xenon (Xe)

7. Light energy per unit area can be defined as _____, and this light energy per unit area per unit time can be again expressed as _____.
   a. Fluence rate, fluence
   b. Fluence, fluence rate
   c. Power, fluence rate
   d. Fluence, power

8. The UV radiation's photochemical effect damages the DNA of the microorganisms, which carries their genetic information, and finally _____ the microorganism.
   a. Kill
   b. Disintegrate
   c. Inactivate
   d. Split

9. The loss of nutrients and enhancement of specific bioactive compounds in the treated food products by UV or PL can be attributed to _____ and _____, respectively.
   a. Photo-oxidation, depolymerization
   b. Depolymerization, photo-oxidation
   c. Photo-reduction, photo-reactivation
   d. Photo-reactivation, photo-reduction

10. According to Beer-Lambert's law, the absorbance (A) of light passing through a medium depends on two parameters other

than the extinction coefficient (ε) of the medium. In this sense which one of the following is true?

a. A is inversely proportional to the concentration of the radiation absorbing species (C) and path length of the light ray (d)

b. A is directly proportional to the concentration of the radiation absorbing species (C) and inversely proportional to the path length of the light ray (d)

c. A is directly proportional to the concentration of the radiation absorbing species (C) and path length of the light ray (d)

d. A is inversely proportional to the concentration of the radiation absorbing species (C) and directly proportional to the path length of the light ray (d)

4.10 Short Answer Type Questions

a. How do UV and PL lamps function and emit light?

b. Explain mathematical relations of Beer-Lambert law and Plank-Einstein's equation?

c. Define fluence, fluence rate, pulse width, and peak power in UV or PL treatment.

d. Define clonogenic death for microorganisms. What is the significance of the photophysical effect in PL-based inactivation of microorganisms?

e. How does the temperature rise in food samples occur due to PL treatment, and how can it be minimized?

4.11 Descriptive Questions

a. Differentiate between PL and UV technology while applied in food.

b. Discuss the microbial inactivation mechanisms of UV and PL processing.

c. Describe the enzymes inactivation mechanism and nutritional changes by UV and PL.

d. How do the type of food product, target microorganism, and relative position of the food sample affect the efficacy of UV or PL processing?

e. What are the challenges associated with UV and PL processing of food products, and how can they be managed?

4.12 Numerical Problems

1) *E. coli* in a vegetable product shows log-linear and tail inactivation model in a particular PL treatment given below.

$$\log N = \log\left[\left(10^{\log N_0} - 10^{\log N_{res}}\right) \times e^{-k_{max}F} + 10^{\log N_{res}}\right]$$

Where N_{res} is the minimum achievable microbial load in that sample and k_{max} (in cm^2/J) is the inactivation rate for unit fluence dose to achieve certain microbial log reduction. If for a dose of 7.2 J/cm^2, log N_0 CFU/g (initial load) and log N_{res} CFU/g (residual load responsible for tailing effect) is 4.56 and 2.82, respectively, the survival count is 660 CFU/g. Calculate the dose required to achieve a 90% reduction for the same inactivation rate.

2) In the sample of liquid buffer solution, *E. coli* showed a log-linear inactivation model in which the logarithmic destruction ratio was directly proportional to fluence of the UV. Total 1 kJ light energy was exposed over 100 cm^2 area, and a 50% reduction in microbial load was achieved.

    a. Calculate the inactivation rate constant (k)
    b. If the sample depth (x) is 1 cm and the sample was exposed for
       60 s, the incident fluence rate (E_o) decreases by 60%. Estimate
       the penetration depth of UV in the solution

References

Caminiti, I. M., Noci, F., Morgan, D. J., Cronin, D. A., & Lyng, J. G. (2012). The effect of pulsed electric fields, ultraviolet light or high intensity light pulses in combination with manothermosonication on selected physico-chemical and sensory attributes of an orange and carrot juice blend. *Food and Bioproducts Processing, 90*(3), 442–448.

Suggested Readings

Bhalerao, P. P., Dhar, R., & Chakraborty, S. (2021). Pulsed light technology in food processing. In K. K. Dash, & S. Chakraborty (Eds.), *Advances in Thermal and Non-thermal Food Process Technology*. CRC Press, Taylor and Francis Group, Boca Raton, Florida, USA. pp. 91–121.

Dhar, R., Basak, S., & Chakraborty, S. (2021). Pasteurization of fruit juices by pulsed light treatment: A review on the microbial safety, enzymatic stability, and kinetic approach to process design. *Comprehensive Reviews in Food Science and Food Safety, 21*(1), 499–540.

Gómez-López, V. M., Koutchma, T., & Linden, K. (2012). Ultraviolet and pulsed light processing of fluid foods. In *Novel Thermal and Non-thermal Technologies for Fluid Foods*. Academic Press, Elsevier, San Diego, California, USA. pp. 185–223.

Koutchma, T. (2009). Advances in ultraviolet light technology for non-thermal processing of liquid foods. *Food and Bioprocess Technology*, 2(2), 138–155.

Koutchma, T., Forney, L. J., & Moraru, C. I. (2019). *Ultraviolet Light in Food Technology: Principles and Applications* (Vol. 2). CRC Press, Taylor and Francis Group, Boca Raton, Florida, USA.

Mahendran, R., Ramanan, K. R., Barba, F. J., Lorenzo, J. M., López-Fernández, O., Munekata, P. E., Roohinejad, S., Sant'Ana, A. S., & Tiwari, B. K. (2019). Recent advances in the application of pulsed light processing for improving food safety and increasing shelf life. *Trends in Food Science & Technology*, 88, 67–79.

Answers for MCQs (sec. 4.9)

1	2	3	4	5	6	7	8	9	10
b	c	a	c	d	d	b	c	a	c

5

Power Ultrasound Processing

5.1 Principles

Ultrasounds are sound vibrations that require a medium to travel. Ultrasound waves have more than 16 kHz frequency, beyond the human audible range (**Fig. 5.1**). The human hearing sound frequency ranges are 20 Hz to 20 kHz. Elephants can hear sound frequencies as low as 14–16 Hz (called infrasonic), and bats can hear as high as 90–200 kHz. In this sense, ultrasound operates from the sonic frequency range.

Like sound (longitudinal waves), ultrasound waves propagate through a fluid medium with a series of compression and rarefaction (**Fig. 5.2**). When the forces exerted by the rarefaction cycle become more significant than the attractive forces (cavitation threshold) between molecules of the fluid, the subsequent formation of voids in the form of cavitation bubbles occurs. These bubbles grow in size over multiple cycles of compressions and rarefactions by a process called rectified diffusion, during which small amounts of vapor or dissolved gas from the surrounding fluid enters into the bubble. Such bubbles, which even survived the compression cycles during their expansion phase, remain stable until a specific equilibrium size is reached as per the applied frequency. The bubbles became unstable beyond their equilibrium size and were influenced by the presence of other neighboring bubbles, causing them to explode and create a localized hotspot with high temperature and pressure of about 4000 K and 1000 atm, respectively (**Fig. 5.2**).

The energy released by the collapse of many cavitation bubbles is responsible for bringing physical and chemical changes in the fluid medium. The approximate size of the bubbles can be expressed in the form of Eq. 5.1. The f and r represent the frequency (MHz) and radius (μm), respectively.

$$f \cdot r = 3 \qquad (5.1)$$

DOI: 10.1201/9781003199809-5

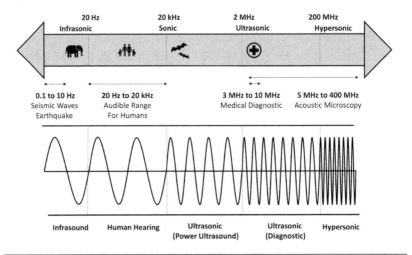

Figure 5.1 Sound spectrum distribution.

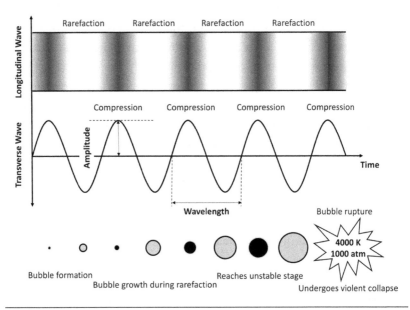

Figure 5.2 Analogy between longitudinal and transverse waves and the propagation of bubbles (formation, growth, and rupture) in an acoustic field.

The collapse of cavitation bubbles formed in the bulk liquid, away from the surface where the liquid evenly surrounds it, can majorly lead to the following three things:

a. Generation of mechanical effects like breakage of polymer chains due to formations shear forces.

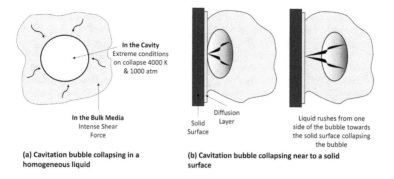

Figure 5.3 Collapse of cavitation bubble (a) within the bulk of a liquid and (b) near solid surface. (Adapted from Mason et al. (2010))

b. Extreme temperature and pressure can break chemical bonds (**Fig. 5.3**).

c. Formation of short-lived hydroxyl radical species from bubbles carrying high energy.

Higher energy is required to initiate the cavitation phenomenon in pure liquids. However, most liquids carry impurities in the form of foreign components and dissolve gases, which provide weak spots that promote the formation of cavitation bubbles at lower energies. Bubbles entrapping gases grow by moving together and merging, quickly rising to the surface. This helps in degassing and improves sonication efficiency by removing air responsible for absorbing acoustic energy (the physical sound waves). When the bubbles collapse on or near the surface, they can cause shock waves targeting the solid surface and facilitate surface cleaning, which is more profound in solid-liquid systems (**Fig. 5.3**). Ultrasound treatment of solid food (probably in bath sonication mode) like meat can also help in improving tenderness, cohesiveness, and water-binding capacity. The use of ultrasound is problematic in the case of gaseous systems due to higher attenuation (loss of power) in the propagation of sound waves.

5.2 Mechanism of Quality Changes in Food

5.2.1 Microbial Inactivation

Ultrasound treatment with a high-acoustic power density (*APD*, W/m³), also known as power ultrasound, can inactivate microorganisms. Typically, power ultrasound possesses high amplitude ultrasonic

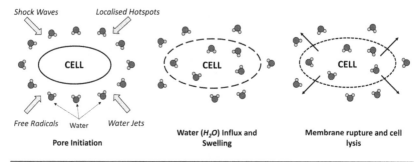

Figure 5.4 Ultrasound based microbial inactivation mechanism.

waves and utilizes acoustic frequencies between 20 and 100 kHz. Opposite to this, ultrasound with low *APD* can be used in bioreactors and fermenters to improve mixing and facilitate microbial growth. There are several hypotheses regarding mechanisms associated with microbial inactivation due to ultrasound. The explosion of transient cavitation bubbles (short-lived bubbles) can produce shock waves, localized hot spots, water jets (only at the solid-liquid interface), and free radicals.

On the other hand, the explosion of stable cavitation bubbles can generate micro-streaming with high shear. All these phenomena are responsible for cell wall and membrane damage, causing cell death (**Fig. 5.4**). As microbial death or inactivation is mainly caused by effects other than heat, ultrasound is still considered a non-thermal processing technology. However, a longer treatment time can result in a temperature rise of the overall sample body. Unlike other non-thermal technologies where sub-lethally damaged (damaged incompletely) cells can heal themselves under favorable conditions (suitable pH, water activity, temperature, and nutrients), cells ruptured or disintegrated by ultrasound cannot recover. Various microorganisms have different resistance toward ultrasound treatment. It has been reported that spores are most resistant, followed by fungi, yeast, Gram-positive bacteria, and, finally, Gram-negative bacteria are the most susceptible microbial entity toward ultrasound treatment. Within the same microorganism, distinct strains can have different resistance.

5.2.2 Inactivation of Spoilage Enzymes

Enzymes are fundamentally complex protein structures. Extreme conditions in terms of localized high pressure and temperature (physical effects) originated from the collapse of cavitation bubbles can break hydrogen bonds and van der Waals forces of interaction present in polypeptide chains of the protein structure. The hydroxyl and free radicals (chemical effects) produced during ultrasound cavitation may also react with amino acid residues of the enzyme's active site and alter its properties. The physical and chemical effects of ultrasound cavitation can render the enzyme inactive. However, it has been observed that the enzyme inactivation takes a relatively long time with ultrasound alone. To speed up the inactivation process and take advantage of the ultrasounds' non-thermal characteristics, it is combined with other non-thermal or mild thermal processing technologies simultaneously or sequentially. For example, treatment with ultrasound, mild heat (~60°C), and low pressure (~400 kPa) for about 30 s can inactivate the quality deteriorating enzyme, pectin methylesterase (PME) activity by 95% in orange juice.

5.2.3 Effect on Nutritional Quality

The nutritional profile of a food material can be affected by ultrasound processing. The formation of free radicals like hydroxyl species during ultrasound cavitation can degrade various nutrients, including vitamin C, anthocyanin, and antioxidants. In addition to that, a longer processing time can cause nutrient loss due to temperature rise. Pyrolysis assisted by the cavitation phenomenon is the primary pathway in the degradation of polar nutrient components. It has also been observed that whole solid food material treated in bath sonication (ultrasound treatment while immersed in water or a suitable hypotonic solution) may cause leach-out of soluble solids and decrease the total soluble solids (TSS) of the food product.

5.3 Equipment and Its Operation

Ultrasound-producing assembly is equipped with electrical devices, known as a transducer, which converts one form of energy into another. There are many types of transducers, out of which four are commonly used for ultrasound: (1) mechanical or liquid-driven, (2) magnetostrictive, (3) electromagnetic, and (4) piezoelectric transducer. Mechanical transducers do not require electrical energy;

in the case of a whistle reactor, a pressurized fluid stream is impacted on a thin blade, which causes it to vibrate at its natural frequency and produce ultrasound. Similarly, in the case of a siren, the fluid flow is interrupted by revolving holes on the rotor, generating fluid puffs at a specific frequency. Such mechanical transducers can produce ultrasound with high amplitudes. Magnetostrictive transducers use ferromagnetic materials (such as iron, nickel, cobalt, certain alloys, and specific ceramic materials), which can change their shape or dimensions under the effect of magnetic fields. Conversely, if its shape is changed due to external perturbation, then the material's magnetic properties also change. Repeatedly and rapidly switching on and switching off the current produces a magnetic field and ferromagnetic materials generate sound vibrations in the presence of this magnetic field.

Electromagnetic transducers utilize the interaction between the magnetic field of a permanent magnet and alternating current moving through a coil to transform electric oscillation into ultrasonic vibrations. Piezoelectric transducers are the most popular devices used to produce ultrasonic vibrations. The active component is a polarized material, such as quartz (SiO_2) or barium titanate ($BaTiO_3$), whose dimensions change under the application of electric field (electrostrictive property). Conversely, it generates an electric charge proportional to the applied stress (Piezoelectric property).

Several ultrasonic devices are equipped with a specific transducer, such as an ultrasonic probe, ultrasonic bath, radial or parallel vibrating system, etc. An ultrasonic bath (**Figs. 5.5a** and **5.6a**) is a metal tank filled with a suitable liquid medium (usually water), which helps in the uniform distribution of sound vibrations. They consist of transducers attached to the tank walls. The power associated with bath-type devices is low to protect the tank from cavitation damage, and the *APD* is also low due to relatively large volume. The ultrasonic probes (**Figs. 5.5b** and **5.6b**) consist of one or more metal horns connected to the transducer (majorly piezoelectric). They can produce high power (≤ 16 kW) and high *APD*, provided that the horn must be designed to resonate at the same frequency as the transducer.

Some of the mathematical terms associated with ultrasonic devices are acoustic power dissipated by the transducer (P_t, W) (Eq. 5.2),

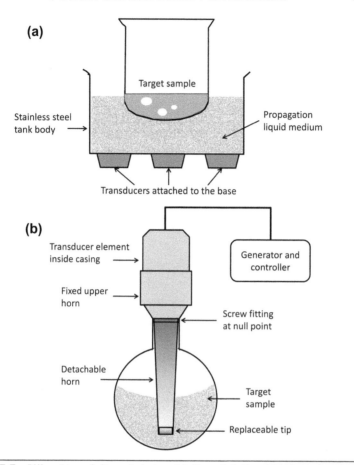

Figure 5.5 Different type of ultrasonic devices: (a) ultrasonic bath and (b) ultrasonic probe.

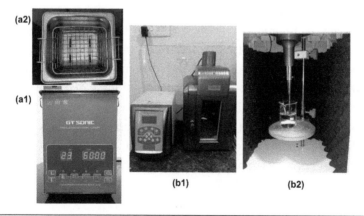

Figure 5.6 Two basic ultrasonic devices in laboratory: (a) bath type sonicator with 'a1' as the front and 'a2' as the top view, and (b) probe type sonicator with 'b1' as the outside and 'b2' as the inside view of the treatment zone of the equipment.

power absorbed (P_a, W) by the medium (Eq. 5.3), acoustic intensity (AI, W/m²) (Eq. 5.4), and APD (W/m³) (Eq. 5.5).

$$P_t = \frac{k_r f^2 A^2}{t} \tag{5.2}$$

$$P_a = mCp \frac{\Delta T}{\Delta t} \tag{5.3}$$

$$AI = \frac{4P_a}{\pi d^2} \tag{5.4}$$

$$APD = \frac{P_a}{V_o} \tag{5.5}$$

In the above equations, k_r is the constant obtained as the product of the resistive dissipation constant, which includes the general oscillator's inertia and other values. Subsequently, A is the sound wave amplitude, t is time, m is mass of the medium, Cp is the specific heat of the medium, ΔT is temperature rise observed in Δt period, d is the internal diameter of the metal horn, and V_o is the volume of the medium under process.

5.4 Critical Processing Factors

5.4.1 Type of Medium

During ultrasound treatment, the acidic environment majorly influences the enzyme inactivation. Moreover, lower water activity can diminish the efficacy of microbial reduction due to ultrasound. For example, 4 log cycle inactivation of *Salmonella eastbourne* in peptone water was achieved in 10 min, but only 0.8 log cycle inactivation was achieved in chocolate after 30 min. Unlike model solutions added with external enzyme or microbial strain, the components in a natural food matrix tend to reduce the lethality of ultrasonic vibrations and show a shielding effect. The sound waves attenuate more in thick medium, and thus higher power is required to achieve the cavitation effect of ultrasound.

5.4.2 Target Microorganism and Enzyme

Microorganisms and enzymes may have variable resistance toward ultrasound treatment. It is known that microorganisms having larger surface areas experiencing the cavitation effect receive more significant damage. Therefore, microbial cells in cocci and other smaller shapes are more resistant than rod-shaped ones. Gram-negative cells have a thinner peptidoglycan layer and overall cell wall than Gram-positive cells. The susceptibility of Gram-negative bacteria (*Escherichia coli* and *Salmonella* spp.) is the highest toward ultrasound, Gram-positive bacteria (such as *Listeria* and *Bacillus* spp.) have moderate susceptibility, and fungal spores (such as spores of *Saccharomyces cerevisiae*) or bacterial spores (such as spores of *Bacillus* spp. and *Clostridium* spp.) are the most resistant. Anaerobic bacteria are less resistant than aerobic ones. Different enzymes or the same enzyme from different sources may show different susceptibility. Ultrasound alone is not much effective for enzyme inactivation (at least 90% reduction in activity) and thus takes a long time to achieve it. The initial microbial load or concentrations of enzymes in the product is also an essential factor during the inactivation process.

5.4.3 *Equipment and Mode of Operation*

The lethality of ultrasound treatment is associated with the cavitation intensity that depends on many physical parameters, such as amplitude, frequency, and power. The size of the cavitation bubbles depends on the frequency of the ultrasonic waves, and more giant bubbles dissipate more energy on collapsing. The majority of the microbial inactivation studies involve the usage of a constant frequency of about 20–24 kHz. However, at a higher frequency of the ultrasound waves, the formation of cavitation bubbles becomes more complex, and no cavitation occurs beyond 2.5 MHz. In terms of the amplitude of the sound waves, higher amplitudes affect the cavitation intensity (*AI*) and tend to produce more cavitation bubbles, which will enhance the inactivation of microorganisms and spoilage enzymes. In many of the reported research works, the relative count or *D*-value (time required to kill 1 log cycle or 90% of the initial count) of the microorganisms has been modelled with process parameters, such as amplitude (*A*, Eq. 5.6) and *APD* (Eq. 5.7).

$$\log D_{US} = \log D_o - 0.0091 \times (A - 62) \tag{5.6}$$

Where D_{US} is the decimal reduction time (min) under ultrasound treatment at constant amplitude and temperature, D_o is the decimal reduction time (min) for ultrasound processing at an amplitude (A) of 62 μm in Eq. 5.6, which apply for the amplitude range of 62–150 μm. In a different expression related to the APD in W/m³ (Eq. 5.7), the D'_{US} is the same decimal reduction time, N_o is the initial microbial count and N is the count after a treatment of t time (min) at temperature T (K), T_{ref} is the reference temperature (K), and the rest of the alphabets from a to d are constants.

$$D'_{US} = a - b \times \ln(APD)$$

$$\ln\left(\frac{N}{N_o}\right) = \frac{1}{D'_{US}} \times t \times \exp\left(\frac{T - T_{ref}}{c - d \times \ln(APD)}\right) \qquad (5.7)$$

5.5 Practical Example

We will consider one ultrasound-based research work as a practical example. In a study regarding ultrasound treatment of apple juice, the effect of processing has been recorded for an inoculated microorganism *Alicyclobacillus acidoterrestris* (Yuan et al., 2009). This is a Gram-positive thermoacidophilic bacteria capable of spore-forming and is responsible for spoilage in juices. It can grow in a pH range of 2.5–6.0, and temperatures range of 20–60°C. *A. acidoterrestris* can survive in acidic medium of juices and high processing temperatures. In this study, bacterial counts were quantified as colony forming unit (CFU) per mL of juice taken as inoculum for the microbial growth medium (plated agar), and a few quality attributes including total sugar content, titratable acidity, turbidity, and brown were measured.

Ultrasound treatment condition: The device operated at a frequency of 20–24 kHz, supply power of 200–700 W, treatment time of 10–60 min, for 100 mL inoculated juice. The study also considered treatments with different initial cell counts (10^4, 10^5, and 10^6 CFU/mL).

Observations after ultrasonic treatment: A more significant impact of ultrasound processing was observed on *A. acidoterrestris* for a higher power supply and longer treatment time. After processing the apple juice at 300 W for 30 and 60 min, a corresponding reduction of 60% and 90% in the microbial count was discovered. The lowest *D*-value was measured at 600 W power and 36.2 min processing time. No significant change was observed in the juice quality.

5.6 Challenges

Ultrasound is considered a clean non-thermal technology with great potential for food processing, preservation, and extraction techniques. Low power ultrasound is also used in fermenters to promote mixing in the media and facilitate the growth of desirable microorganisms. Power ultrasound can inactivate undesirable microorganisms with less or negligible food quality and sensory attributes. Still, it is ineffective against enzymes and few resistant microorganisms (such as yeasts and molds) and takes a long time to inactivate them. So, to overcome such limitations, ultrasound needs to be combined with other non-thermal or mild thermal technologies. Unlike other non-thermal processing technologies, sub-lethally damaged microorganisms cannot repair themselves after ultrasound treatment. Even though ultrasound is a non-thermal technique, a longer processing time can increase product temperature, which can be controlled by incorporating specific cooling mechanisms or reducing treatment time by combining with other technologies or hurdles. In the case of probe sonicators, sometimes there is a chance of food contamination from the corrosion of the metallic horn. However, noncontact type ultrasound devices also exist. The majority of the ultrasonic devices are comparatively easy to operate and are less costly than other non-thermal processing equipment.

5.7 Summary

- Ultrasounds are sound waves with a frequency of more than 16 kHz, propagating through a medium with a series of compression and rarefaction.
- Ultrasound is one of the non-thermal food processing technologies that use sound waves to inactivate microorganisms with minimum effect on the nutrients.
- Power ultrasound possesses high amplitude ultrasonic waves and utilizes 20 and 100 kHz acoustic frequencies.
- Cavitation bubbles are born from the formation of voids that developed when the forces exerted by the rarefaction cycle becomes more significant than the cavitation threshold of the fluid.

- The collapse of cavitation bubbles can locally generate high mechanical shear forces, high temperature, and the formation of short-lived hydroxyl radical species.
- Localized shock waves, high temperature, and free radicals can inactivate microorganisms through the cell wall and membrane damage.
- Sub-lethally injured microorganisms after ultrasound treatment cannot recover.
- With ultrasound, it is difficult to inactivate spoilage enzymes and microbial spores and thus requires a very long treatment time.
- Ultrasound can improve the extractability of soluble nutrients or bioactive compounds; however, prolonged treatment can degrade them due to temperature rise and reaction with free radicals.
- The four commonly used transducers for ultrasound are mechanical or liquid-driven, magnetostrictive, electromagnetic, and piezoelectric transducer.
- Mechanical transducers do not require electrical energy such as whistle reactor.

5.8 Solved Numerical

1) Ultrasound technology (probe-type) has been used to treat 100 mL orange juice for 10 min. During treatment, the amplitude was fixed at 110 μm. It was observed that the radius of the average cavitation bubbles was 300 μm. The probe transducer's overall resistive dissipation constant (k_r) is 13.56 kJ$\cdot$Hz$^{-1}\cdot$m^{-1}. Assume that about 30% of the power dissipated by the device has been absorbed by the sample medium, which got converted into heat. Consider the specific heat (Cp) and density (ρ) of the orange juice as 3.89 kJ$\cdot$kg$^{-1}\cdot$°C^{-1} and 1150 kg$\cdot$m^{-3}, respectively. Calculate the possible temperature rise (ΔT) in the medium and the *APD* absorbed by the medium.

Solution: Given information:

Treatment time (t) = 10 min = 600 s
Ultrasonic wave amplitude (A) = 110 μm = 110 $\times$ 10^{-6} m
Cavitation bubble radius (r) = 300 μm
Overall resistive dissipation constant of the probe transducer (k)
= 13.56 kJ$\cdot$Hz$^{-1}\cdot$m^{-1}

Specific heat of the orange juice (Cp) = 3.89 kJ·kg⁻¹·°C⁻¹

Juice density (ρ) = 1150 kg·m⁻³

Juice treatment volume (V_o) = 100 mL = 1×10^{-4} m³

Calculating mass of treated juice $(m) = V_o \times \rho = (1 \times 10^{-4}) \times 1150 =$ 0.115 kg

Frequency (f, MHz) can be related to radius of the cavitation bubble $(r, \mu m)$; $f \times r = 3$ => $f \times 300 = 3$ => $f = 0.01$ MHz = 10 kHz

Now, we can calculate the power dissipated (P_t) by the transducer;

$$P_t = \frac{k_r \cdot f^2 \cdot A^2}{t} \quad \Rightarrow P_t = \frac{13.56 \times (10 \times 10^3)^2 \times (110 \times 10^{-6})^2}{600}$$

$$= 0.02735\ \text{kW} = 27.35\ \text{W}$$

30% of the P_t is converted into heat. So, the power absorbed by the medium, $P_a = 0.3 \times P_t => P_a = 0.3 \times 27.35 = 8.205$ W

Calculating for ΔT; $P_a = m \cdot Cp \cdot \dfrac{\Delta T}{\Delta t}$

$$\Rightarrow 8.205 = 0.115 \times 3.89 \times 1000 \times \frac{\Delta T}{600} \quad \Rightarrow \Delta T = 11.0°C$$

$$APD = \frac{P_a}{V_a} \quad \Rightarrow APD = \frac{8.205}{100} = 0.08205\ \text{W·mL}^{-1}$$

2) A buffer solution inoculated with a strain of *Salmonella enterica* has been treated with ultrasound. An experiment has been conducted for the buffer solution at power density ranging from 783 to 1200 W/L, and decimal reduction time (D) of the microorganism was noted. Refer to the following table showing the observed effect of ultrasound on the D values of *S. enterica* at different power densities (APD).

APD (W·L⁻¹)	D (min)
783	0.78
383	1.14
165	2.16
103	2.44
68	2.98

A logarithmic relation has been developed to predict the D values. The model constants a and b are −0.918 and 6.778, respectively.

$$D_p = a \times \ln(APD) + b$$

Calculate the absolute average error ($|e|_{avg}$) in the prediction of D values by the model.

Solution:

  a. At $APD = 783$ W·L^{-1}, $D_{p1} = -0.918 \times \ln{(783)} + 6.778 =$
  0.66 min
  b. At $APD = 383$ W·L^{-1}, $D_{p2} = -0.918 \times \ln{(383)} + 6.778 =$
  1.32 min
  c. At $APD = 165$ W·L^{-1}, $D_{p3} = -0.918 \times \ln{(165)} + 6.778 =$
  2.09 min
  d. At $APD = 103$ W·L^{-1}, $D_{p4} = -0.918 \times \ln{(103)} + 6.778 =$
  2.52 min
  e. At $APD = 68$ W·L^{-1}, $D_{p5} = -0.918 \times \ln{(68)} + 6.778 =$
  2.90 min

Absolute (positive) error, $e = |D - D_p|$

$$e_1 = |0.78 - 0.66| = 0.12$$

$$e_2 = |1.14 - 1.32| = 0.18$$

$$e_3 = |2.16 - 2.09| = 0.07$$

$$e_4 = |2.44 - 2.52| = 0.08$$

$$e_5 = |2.98 - 2.90| = 0.08$$

Average absolute error,

$$|e|_{avg} = \frac{e_1 + e_2 + e_3 + e_4 + e_5}{5} = \frac{0.12 + 0.18 + 0.07 + 0.08 + 0.08}{5}$$

$$\Rightarrow |e|_{avg} = 0.10$$

3) Sequential effect of heat and ultrasound (thermosonication) has been utilized to inactivate PME in tomato juice. The food sample was preheated to the required temperature (50–75°C) and immediately treated with ultrasound. During sonication, the temperature was maintained at 3°C with the help of a recirculating water bath. The inactivation of the spoilage enzyme PME was observed to follow the first-order kinetics. The details of the experiment have been given in the following table.

S.NO.	PREHEATING TEMPERATURE (°C)	SONICATION TIME (t, min)	OVERALL INACTIVATION [$\ln(A/A_0)$]
1	50	10	−0.5
2	50	75	−1.9
3	75	2	−0.9
4	75	5	−5.3

A, enzyme activity after treatment for at time t; A_0, initial enzyme activity.

Calculate the D-value (decimal reduction time), z-value (increase in temperature required to achieve 10 fold (1 $\log_{10}$) reduction in D-value), and the activation energy (E_A) for PME inactivation by thermosonication.

Solution: The inactivation curve for PME follows first-order kinetics with k (min^{-1}) as the inactivation rate constant;

$$\ln\left(\frac{A}{A_0}\right) = -k \cdot t$$

For single-temperature, k remains constant. So, we will get two k-values for two temperatures. We use a straight line's equation ($y = mx$) to solve the kinetic equation for obtaining a mean or overall k (negative slope representing inactivation) for each temperature. Here, $\ln(A/A_0)$ is the y-axis, t is the x-axis, and $-k$ is the slope (m).

$$m = \frac{y_2 - y_1}{x_2 - x_1} \qquad -k = \frac{\left|\ln\left(\frac{A}{A_0}\right)\right|_2 - \left|\ln\left(\frac{A}{A_0}\right)\right|_1}{t_2 - t_1}$$

a. At 50°C,

$$-k_1 = \frac{-1.9 - (-0.5)}{75 - 10} \qquad => k_1 = 0.022 \text{ min}^{-1}$$

b. At 75°C,

$$-k_2 = \frac{-5.3 - (-0.9)}{5 - 2} \qquad => k_2 = 1.467 \text{ min}^{-1}$$

Calculating D-value: $D = 2.303/k$

a. At 50°C, $D_1 = 2.303/0.022 = 106.93$ min
b. At 75°C, $D_2 = 2.303/1.467 = 1.57$ min

It can be observed that with high-temperature decimal reduction time decreases.

Now calculating z-value;

$$\log D_{ref} - \log D = \frac{T - T_{ref}}{z} \quad => z = \frac{T - T_{ref}}{\log\left(\dfrac{D_{ref}}{D}\right)}.$$

Considering treatment condition of 50°C as the reference.

$$z = \frac{75 - 50}{\log\left(\dfrac{106.93}{1.57}\right)} \quad => z = 13.64°C$$

Finally, calculating E_A value using linearized Arrhenius equation;

$$\ln k = \ln k_{ref} - \frac{E_A}{R}\left(\frac{1}{T} - \frac{1}{T_{ref}}\right).$$

Here, R (gas constant) = 0.008314 kJ·mol^{-1}·K^{-1} and temperature is in Kelvin (K). Also, considering treatment condition of 50°C as the reference.

$$E_A = \ln\left(\frac{k_{ref}}{k}\right) \times R \times \frac{T \cdot T_{ref}}{(T_{ref} - T)}$$

$$=> E_A = \ln\left(\frac{0.022}{1.467}\right) \times 0.008314 \times \frac{348 \times 323}{(323 - 348)}$$

$$=> E_A = 157.75 \text{ kJ} \cdot \text{mol}^{-1} \cdot \text{K}^{-1}$$

The z-value and E_A are properties specific to the PME in tomato juice of a specific variety, and hence a single value is obtained for them.

5.9 Multiple Choice Questions

1. Ultrasound wave comes under the category of _____ waves, and during their propagation, the particles of the medium vibrate _____ to the direction of the wave propagation.
   a. Transverse, perpendicular
   b. Transverse, parallel

    c. Longitudinal, parallel
    d. Longitudinal, perpendicular

2. Microorganisms are inactivated by ultrasound treatment because of membrane damage mainly due to:
    a. Localized high-pressure spots
    b. Localized high-temperature spots
    c. Free radicals
    d. All of them

3. The fundamental parameters of ultrasound, mainly responsible for 'size of the cavitation bubble' and the 'AI of cavitation' or 'number of bubbles,' respectively, are _____ and _____.
    a. Supplied electric power, wave frequency
    b. Wave frequency, wave amplitude
    c. Wave amplitude, supplied electric power
    d. Supplied electric power, wave frequency

4. Piezoelectric ultrasonic transducers consist of _____ as its core or active component.
    a. Polarized material (like quartz)
    b. Ferromagnetic material (like iron)
    c. Permanent magnet
    d. Copper coil with AC current running through it

5. The category of microorganisms that offer the least and highest resistance toward ultrasound processing are _____, respectively.
    a. Gram-negative bacteria and Gram-positive bacteria
    b. Gram-positive bacteria and microbial spores
    c. Gram-negative bacteria and microbial spores
    d. Microbial spores and Gram-positive bacteria

6. These bubbles grow in size over multiple cycles of compressions and rarefactions by a process called _____, during which small amounts of vapor or dissolved gas from the surrounding fluid enter the bubble.
    a. Bubble expansion
    b. Rectified diffusion
    c. Unrectified diffusion
    d. Vapor infusion

7. The formation of cavitation bubbles is difficult for _____.
   a. Transparent solution
   b. Isotonic solution
   c. Impure solution
   d. Pure solution

8. Can the sub-lethally damaged (damaged incompletely) microbial cells heal themselves after exposure to ultrasonic waves?
   a. No
   b. Yes
   c. Yes, they may heal only under favorable conditions
   d. Not sure; need further research for confirmation

9. Generation of extreme conditions due to the collapse of cavitation bubbles; (1) can break _____ present in polypeptide chains of the protein structure because of localized high pressure and temperature (physical effects), and (2) the hydroxyl and free radicals produced during ultrasound cavitation can react with _____ of the enzyme and alter its properties (chemical effects).
   a. 'Amino acid residues or active site', 'hydrogen bonds and van der Waals forces of interaction'
   b. 'Primary bonds', 'secondary structures'
   c. 'Hydrogen bonds and van der Waals forces of interaction', 'amino acid residues or active site'
   d. 'Secondary structures', 'primary bonds'

10. To generate ultrasonic vibrations, magnetostrictive transducers can change their shape or dimensions under the effect of _____ to produce ultrasound. Whereas the electromagnetic transducers simply utilize _____ to produce ultrasonic vibrations.
    a. 'Ferromagnetic materials in the presence of the magnetic field', 'permanent magnet and electric field'
    b. 'Permanent magnet and electric field', 'ferromagnetic materials in the presence of magnetic field'
    c. 'Ferromagnetic materials in the presence of the electric field', 'permanent magnet and magnetic field'
    d. 'Permanent magnet and magnetic field', 'ferromagnetic materials in the presence of electric field'

5.10 Short Answer Type Questions

a. How are bubbles formed and grow during ultrasound treatment of a fluid?

b. What is ultrasonic cavitation, and what effects do they show when cavitation bubbles collapse in the bulk of the fluid?

c. Describe the working principles of different types of ultrasound transducers.

d. With the help of schematics, discuss the working principles of bath and probe-type sonication devices.

e. How is the efficacy of ultrasound treatment affected by the type of sample medium?

5.11 Descriptive Questions

a. Explain the susceptibility and mechanism of microbial inactivation during ultrasound processing.

b. How do the ultrasonic waves affect the spoilage enzymes and the nutritional qualities of a food product?

c. How does the mode of operation influence the lethality of ultrasound treatment?

d. Describe the influence of critical process parameters on the cavitation phenomenon during ultrasound treatment of food.

e. Discuss the advantages and limitations of power ultrasound processing of food samples.

5.12 Numerical Problems

1) High-intensity ultrasound processing of pineapple juice has been performed. The equipment is equipped with a metallic horn-type sonication device with a 13-mm diameter titanium probe. The maximum operating power of the equipment is 500 W. About 150 mL food sample was treated at 19 kHz frequency for 10 min at three different power levels; 20%, 60%, and 100% of the maximum operating power. The corresponding temperature rise was 17, 44, and 50°C. Consider the specific heat (Cp) and density (ρ) of the pineapple juice as 3.68 kJ·kg^{-1}·°C^{-1} and 1057 kg·m^{-3}, respectively. Calculate the power absorbed by the food sample (P_a), energy conversion efficiency (η, as per the power delivered to the device and power absorbed by the sample), AI, and APD.

2) Following the same experimental conditions of the previous numerical problem (*Q1*). Some extra response data regarding the effect of ultrasound on the activity of polyphenoloxidase (PPO) enzyme and total phenolic content (TPC) in pineapple juice has been provided in the following table.

S.NO.	EQUIPMENT OPERATION POWER (%)	SONICATION TIME (t, min)	PPO ACTIVITY (%)	TPC $[\mu g \cdot (100\ mg)^{-1}]$
1	Untreated	0	100	10.78
2	20	10	84.42	11.28
3	100	10	80.81	14.00

The operating power of the equipment is expressed as the percentage of the maximum operating power, 500 W.

Develop two modified first-order kinetic models, showing log natural of 'relative enzyme activity' or 'relative phenolic content' as a function of *AI* ($W \cdot cm^{-1}$).

References

Yuan, Y., Hu, Y., Yue, T., Chen, T., & Lo, Y. M. (2009). Effect of ultrasonic treatments on thermoacidophilic *Alicyclobacillus acidoterrestris* in apple juice. *Journal of Food Processing and Preservation*, *33*(3), 370–383.

Suggested Readings

Kentish, S., & Feng, H. (2014). Applications of power ultrasound in food processing. *Annual Review of Food Science and Technology*, 5, 263–284.

Mason, T. J., Paniwnyk, L., Chemat, F., & Vian, M. A. (2011). Ultrasonic Food Processing. In A. Proctor (Ed.), *Alternatives to Conventional Food Processing*. The Royal Society of Chemistry, Milton Road, Cambridge, UK. pp. 387–414.

Rastogi, N. K. (2011). Opportunities and challenges in application of ultrasound in food processing. *Critical Reviews in Food Science and Nutrition*, *51*(8), 705–722.

Villamiel, M., García-Pérez, J. V., Montilla, A., Carcel, J. A., & Benedito, J. (Eds.). (2017). *Ultrasound in Food Processing: Recent Advances*. Wiley-Blackwell, Chichester, West Sussex, UK.

Answers for MCQs (sec. 5.9)

1	2	3	4	5	6	7	8	9	10
c	d	b	a	c	b	d	a	c	a

6

COLD PLASMA PROCESSING

6.1 Principles

Plasma is considered the fourth state of matter, which can be viewed as a gas of ions, electrons, free radicals, atoms, and molecules at the ground or excited state. It can be generated from complete ionization of the surrounding gas due to high temperature (millions of kelvin) like in a nuclear reactor. Subsequently, incomplete ionization can be divided into thermal and non-thermal plasma. Thermal plasma, for example, in the case of natural lighting and welding arc, requires high temperature (1 eV ≈ 11,600 K) or high power (not more than 50 MW), where all the particles (electrons and heavier species, such as ions and molecules) are at thermodynamic equilibrium. This category also includes the quasi-equilibrium plasma associated with a temperature range of 100–150°C. At the same time, non-thermal plasma imparting temperature < 60°C is produced at low power and near atmospheric pressure, without any localized thermodynamic equilibrium among the electrons and heavier species. Cold plasma is also called nonequilibrium or non-thermal plasma. This technology does not create or require cooling conditions; therefore, the term 'cold' must not be misunderstood. Choosing the suitable gas as feed for the system converted into plasma is crucial, depending on the application. It can be a pure gas, such as helium (He), nitrogen (N_2), argon (Ar), a specific mixture of gases, or even simple air with specific moisture.

The plasma consists of a balanced concentration of charged particles resulting in an overall neutral state. Plasma formation is a result of two processes that include: (1) ionization stage and (2) recombination stage (**Fig. 6.1**). The ionization stage is the initial stage of the plasma generation process, for example, the electrical breakdown of gas molecules into various species at excited and ground states (incomplete ionization). Naturally, the generated species at excited states that are unstable will try to stabilize themselves or come to a ground state.

DOI: 10.1201/9781003199809-6

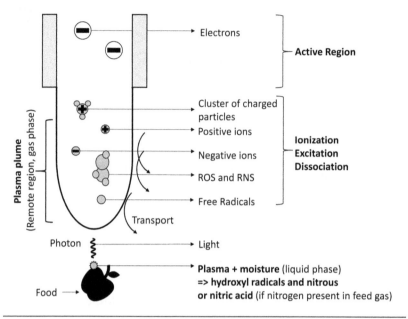

Figure 6.1 Plasma products and their interaction with food sample at both gas and liquid phase (moisture present on sample surface).

While doing so, in the recombination stage, plasma decay occurs, and the energy of initially formed ionized plasma gets dissipated in the form of heat, chemical energy (initiating specific reactions), light (maybe in visible or UV range), etc.

Ionization occurs after removing at least one electron from an atom or a molecule. In majority of cases, ionization is achieved by electron impact, during which an electron (e^-) attaining high kinetic energy under the influence of the electric field of the cold plasma device knocks out the valence electron of an atom (A) or molecule (AB) present in the feed gas (Eq. 6.1).

$$A + e^- \rightarrow A^+ + 2e^- \tag{6.1}$$

It will only work if the colliding electron has enough kinetic energy to overcome the ionization potential of that atom or molecule. Apart from electron impact, other types of ionization techniques can also occur. In the recombination stage, when the positive ion and electron come together to form a neutral atom or molecule, they still possess all the ionization energy present since the ionization stage. Unless the atom or molecule ionizes again, this energy must be dissipated in some other way. Initiation of dissociative recombination (Eq. 6.2) process can be one possible mechanism. An electronically excited intermediate (AB*) is temporarily formed in this pathway and later breaks down into neutral atom 'A' and excited atom 'B*'. Such a mechanism is already used in destroying pollutants using plasma treatment. The symbol "*" refers to an unpaired valence electron.

$$(AB)^+ + e^- \rightarrow (AB)^* \rightarrow A + B^* \tag{6.2}$$

Another common pathway can be radiative recombination (Eq. 6.3). An ion combines with an electron, forming an excited intermediate followed by the release of the photon (having energy $E = h\nu$) in the visible or UV range.

$$A^+ + e^- \rightarrow A^* \rightarrow A + h\nu \tag{6.3}$$

Besides Eqs. 6.2 and 6.3, other recombination mechanisms also exist.

6.2 Mechanisms of Changes in Quality Attributes of Food

Cold plasma inactivates microorganisms mainly by three mechanisms: (1) chemical reaction of the cellular membrane with charged particles like electron and ions, radicals (OH, O, etc.), reactive or excited molecules (NO, O_2^*, O_3, etc.), (2) UV radiation generated during cold plasma treatment can erode cell membrane and damage cellular components, and (3) degradation of DNA by UV (**Fig. 6.2**). Among the reactive oxygen species (ROS), ozone (O_3), atomic oxygen (O), singlet oxygen ($^1O_2$), superoxide (eg. superoxide anion O_2^-), peroxide (e.g., H_2O_2), and hydroxyl radicals ($^•OH$) are found to be essential for microbial inactivation.

Similarly, the reactive species, charge particles, possible UV light present in plasma can inactivate or denature enzyme proteins by reacting with the secondary structure of the protein (causing loss of α-helix and β-sheet), with amino acid residues, and modifying side chains. Through similar pathways, cold plasma may affect nutrients of the food materials, including protein, carbohydrate, vitamins, antioxidants, color

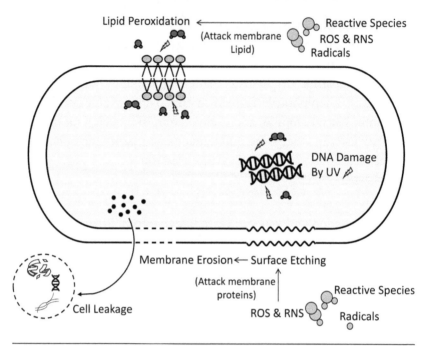

Figure 6.2   Effect of cold plasma on microbial cell.

components. In the case of grains and legumes, it has been reported that cold plasma treated carbohydrates and proteins have developed better solubility in water, causing the overall food material to uptake water more efficiently and become soft and less firm in texture. This can be attributed to the surface etching effect of plasma, which can be helpful during soaking and cooking operations. Plasma may also cause lipid oxidations due to interaction with radicals, reactive nitrogen species (RNS), and ROS. This has been observed in a few fresh meat cases of pork, beef, and during partial hydrogenation of soybean oil.

Cold plasma can degrade pesticides, aflatoxins, and food allergens through chemical reactions with reactive species through dissociative recombination mechanisms, and similar pathways. In many studies, it has been discovered that the pH of the product decreased after treatment. This occurred due to the formation of acids on the surface exposed to the reactive plasma species like the formation of nitric acid due to the reaction of moisture and NO (**Figs. 6.1** and **6.2**).

However, all these effects were not the same for other food products. Several researchers also have reported no significant effect of cold plasma on pH, acidity, texture, antioxidants, color, vitamins, and lipids.

6.3 Equipment and Its Operation

There are various devices capable of producing plasma at atmospheric pressure, which include gliding arc discharge (GAD), dielectric barrier discharge (DBD), corona discharge, radio frequency plasma (RFP), and plasma jet (**Fig. 6.3**). In the gliding arc, the gas (usually humid air) is allowed to pass through between two diverging metal electrodes possessing a potential difference of at least 9 kV and a current of 100 mA (**Fig. 6.3b**). The arc is formed between the electrode's narrowest region, which is easily possible in the open air.

DBD (**Fig. 6.4a**) is the most common cold plasma device called silent discharges. It consists of two parallel electrodes covered with dielectric material (ceramic, quartz, etc.) and placed apart from each other, which can vary from 100 mm to several centimeters. The gas lines in the discharge gap require a supply of alternating current with 10 kV ignition voltages. The device has a relatively simple design, flexible electrode geometry, allows uniform gas discharge, and can use different gas compositions. However, it involves high voltage and direct contact of food products and the dielectric coated electrodes. A corona discharge (**Fig. 6.3a**) is generated when the electric field

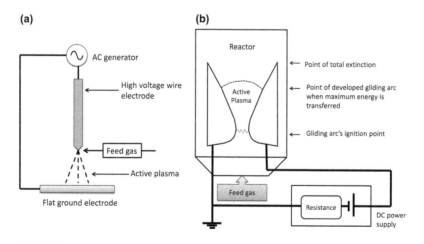

(a)

AC generator

High voltage wire electrode

Feed gas

Active plasma

Flat ground electrode

(b)

Reactor

Active Plasma

Feed gas

← Point of total extinction

← Point of developed gliding arc when maximum energy is transferred

← Gliding arc's ignition point

Resistance

DC power supply

Figure 6.3 Schematic diagram of (a) corona discharge plasma and (b) gliding arc discharge plasma.

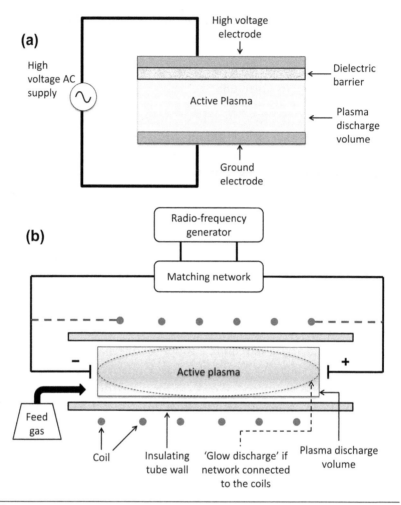

Figure 6.4 Schematic diagram of (a) dielectric barrier discharge plasma and (b) radio-frequency plasma.

exceeds the ionization threshold in a small spatial region. Highly asymmetric electrodes are used as a plane or point. Since the discharge volume/area lie on the end of one of the electrodes, the electric field strength falls below the threshold value, and the plasma is no longer self-sustaining in that region. Thus, diffuse plasma is generated with a low density and blown on the sample. Similar to that, plasma driven by radiofrequency is also blown on the surface of the target product. However, RFP (**Fig. 6.4b**) is generated by placing the feed gas in the presence of oscillating electromagnetic waves of radio frequency originating from different electrodes or induction coils kept outside the reactor region. Food products close to the electrode or cold plasma region get exposed to active plasma containing both short- and long-lived plasma particles and species. When the food product is kept at a certain distance from the electrodes or cold plasma region, it experiences majorly the long-lived plasma species known as quenched or decaying plasma.

A plasma jet is a device in which the feed gas or gas mixture is blown into the discharge region, having the concentric electrodes causing the gas to ionize and form plasma (**Fig. 6.5a**). The inner electrode has a 100–250 V voltage and a current commonly at 13.56 Hz frequency. The plasma is directed out of the nozzle, taking the form of a plasma plume, and pointed to the target food sample. Primarily the long-lived reactive species reaches the sample. The jet source is also called a remote source, and no direct contact of the food with the device takes place.

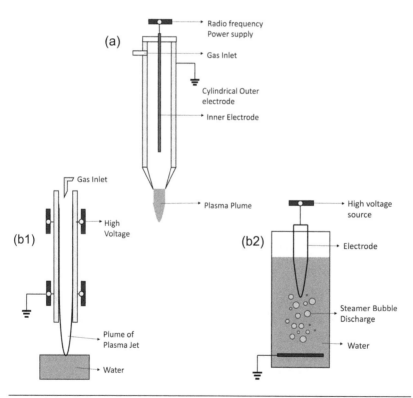

Figure 6.5 Schematic diagram of (a) plasma jet, (b1) plasma-activated water by applying plasma (quenched plasma) on the liquid surface, (b2) plasma-activated water by producing plasma (applying active plasma) within the bulk of the liquid.

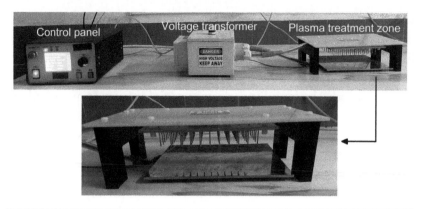

Figure 6.6 Pin to plate type cold plasma setup installed at the Food Engineering and Technology Department, Institute of Chemical Technology, Mumbai, Maharashtra, India.

There are numerous other types of cold plasma devices (including the modified versions of the currently explained devices) available for food processing, such as plasma pencil, plasma needle, diode plasma discharge, microwave powered plasma, and pin to plate type plasma (**Fig. 6.6**). Chemical augmentation of plasma, such as plasma-activated water, has also been explored. The cold plasma's active species are supplemented with hydrogen from moisture (**Fig. 6.5b**). Plasma-activated water is produced by the interaction of plasma with water, which further creates various ROS, RNS, and causes the formation of peroxides like hydrogen peroxide and acids like nitrous and nitric acids (in the presence of nitrogen and oxygen), which show bactericidal effects.

6.4 Critical Processing Parameters

6.4.1 Type of Product

The efficacy of the cold plasma treatment varies for solid and liquid food materials having different properties. Cold plasma works better on the surface and may not penetrate deeper into the product, especially in solids (sufficient literature not available regarding its ability to penetrate). The liquid food products tend to vaporize during treatment. It has been observed that higher microbial inactivation occurred after cold plasma treatment for food materials having lower pH. In addition to that, higher moisture or humidity facilitated better inactivation of microorganisms due to increased formation of hydroxyl radicals.

6.4.2 Target Microorganism or Enzyme

Different microorganisms and enzymes have varying sensitivity toward the cold plasma. Generally, bacteria in the exponential growth phase are more sensitive to any harsh processing environment than the stationary phase. Gram-positive bacteria are more resistant than Gram-negative ones because of their thick peptidoglycan layer and lipopolysaccharide membrane. Similar to other processing technologies, bacterial or fungal spores are more resistant than vegetative ones. Moreover, yeasts and fungi are more resistant than Gram-positive bacteria due to their cell wall reinforced with chitin that protects them from the active species of the cold plasma. A large initial microbial count and high enzyme concentration can decrease the efficacy of the treatment. The first-order kinetics and Weibull distribution model (refer Eq. 3.4) have been commonly used to predict the inactivation trends.

6.4.3 Equipment and Mode of Operation

The significant process variables affecting the efficacy of cold plasma are: composition of the feed gas, supplied voltage, current, frequency range, the relative positioning of the product, and other equipment specifications. The quality and nature of reactive species generated during plasma formation primarily depend on the gas composition and frequency and voltage parameters. The high potential difference corresponds to high energy density exposure for the product. The relative positioning of the food sample from the plasma source is also essential. Sample in contact with the electrode like in DBD experiences the highest energy density and many active plasma species. In the case of GAD, corona discharge, and RFP, the sample usually kept directly in or close to the active plasma zone (close to the electrode but not in contact) receives active plasma species. However, for the same devices, if the sample is placed relatively far from the plasma source, usually during remote treatment, the food product mainly receives the quenched plasma consisting of secondary species (long-lived), which have lower energy than the active zone species. A remote-treatment condition reduces the chance of transferring heat to the food material.

6.5 Practical Example

We will consider one cold plasma-based research work as a practical example for better understanding. The study involves radio-frequency driven cold plasma treatment of groundnuts, separately inoculated with two microorganisms; *Aspergillus parasiticus* and *Aspergillus flavus* (Devi et al., 2017). The plasma effect was also tested on the aflatoxins produced by these microorganisms. *A. parasiticus* and *A. flavus* are mold species capable of producing aflatoxins, which are harmful secondary metabolites of the polyketide class. Aflatoxins fall under mycotoxins, known for their carcinogenic and teratogenic characteristics. In this study, bacterial counts were quantified as colony forming unit (CFU) per g of the sample taken to prepare the inoculum for the microbial growth medium (plated agar).

RFP treatment condition: The device operated at a frequency of 13.6 MHz, supply power of 60 W, and 10 g inoculated groundnut sample was exposed to cold plasma. The cold plasma treatment was conducted in an environment with 24°C and 45.3% relative humidity.

Observations after cold plasma treatment: A more significant impact of cold plasma processing was observed on the microorganisms for higher power supply and longer treatment time. After processing the inoculated groundnuts at 60 W for 24 min, a reduction of 97.9% and 99.3% in the count of *A. flavus* and *A. paraciticus* was discovered, respectively. On the other hand, cold plasma treatment of 60 W for 12 min successfully reduced the aflatoxin-B1 up to 94.3% and 96.8% secreted by *A. paraciticus* and *A. flavus*, respectively.

6.6 Challenges

Cold plasma has shown significant potential as a non-thermal food processing technology. Besides microbial decontamination, this technology also helps modify food components, seed germination, degrade agrochemical residues, and food allergens. However, this technology cannot penetrate solid food materials. It can only partially penetrate liquid food products (depending on its transmittance and turbidity), making it better suited for surface treatments. Treating the food sample in a thin layer form can improve its efficacy. The formation of localized hot streamers may leave some marks on the surface of solid food materials, such as in case of mangoes and melons. High oxidative action from the reactive species from cold plasma can negatively affect bioactive components, proteins, and lipids. Thus, cold plasma may not be suitable for treating high-fat food. The cold plasma device can be very sophisticated in a few cases, such as RFP. Complex machinery and operation may demand trained personnel for handling it properly. During DBD-based cold plasma, the chance of food contamination exists. Suppose the relative placement of the food sample allows it to contact the electrode of DBD that is mostly coated or covered with a dielectric material. One must be careful and actively monitor food sample temperature because, during cold plasma processing with specific devices like plasma jet or microwave powered plasma, sample temperature may exceed 60°C. If the feed gas composition involves pure, noble gases, such as argon, helium, etc., it can make the process more costly. The system needs a very high voltage supply (> 3 kV) that requires extra hardware support and safety measures to avoid accidents. The cold plasma electrodes handle very high levels of current density and, therefore, must be designed and controlled to avoid the formation of arcs (unwanted electric discharge or sparks) on them. Thus, the sample temperature and food contamination will be controlled with a longer electrode life without melting it.

6.7 Summary

- Cold plasma is also called nonequilibrium or non-thermal plasma, which can be produced at low temperature and near atmospheric pressure.
- The plasma consists of a balanced concentration of charged particles resulting in an overall neutral state.
- Plasma formation is a result of two processes that include the ionization stage followed by the recombination stage.
- In the recombination stage, plasma decay occurs, and the energy of initially formed ionized plasma gets dissipated in the form of heat, chemical energy, and light.

- The ionization energy remaining in the second stage or the recombination stage may be dissipated in the form of dissociative or radiative recombination.
- Cold plasma inactivates microorganisms by chemically damaging cell membrane with charged species, free radicals, and UV along with DNA denaturation by UV.
- The reactive species and UV light present in plasma can denature enzyme proteins by damaging the secondary structure, amino acid residues, and by modifying side chains.
- There are various devices capable of producing plasma, such as the GAD, DBD, corona discharge, and radiofrequency plasma.
- Cold plasma works better as a surface disinfectant, and its inability to penetrate deeper into the product is a concern.
- The cold plasma system usually needs a very high voltage supply (> 3 kV) that requires extra hardware support and safety measures to avoid any accidents.

6.8 Solved Numerical

1) Apple juice inoculated with a pathogenic bacteria *Citrobacter freundii* was inactivated up to a certain degree using jet-type cold plasma equipment. The input gas consisted of argon and oxygen. The inactivation curve was fitted to the Weibull model. All the experimental data has been tabulated below.

CONDITION	TREATMENT TIME (t, s)	LOGARITHMIC COUNT (log M)
Control	0	7.7
1	240	6.8
2	360	5.6

Calculate the scale parameter (δ) and shape parameter (β).

Solution: Considering $\log_{10}$ form of the Weibull model,

$$\log (N/N_i) = -(t/\delta)^\beta \quad \Rightarrow \log N_i - \log N = (t/\delta)^\beta$$

Taking log on both sides again, $\log(\log N_i - \log N) = \beta(\log t - \log \delta)$

| Condition 2: | $\log(7.7 - 5.6) = \beta(\log 360 + \cancel{\log \delta})$ |
| Condition 1: | $-\{\log(7.7 - 6.8) = \beta(\log 240 + \cancel{\log \delta})\}$ |

$$0.32 - (-0.05) = \beta(\log 360 - \log 240)$$

$$\Rightarrow 0.37 = \beta(2.56 - 2.38) \qquad \Rightarrow \beta = 2.06$$

Now calculating scale parameter (δ), $\log(7.7 - 5.6) = 2.06(\log 360 - \log \delta)$

$$\Rightarrow 0.32/2.06 = 2.56 - \log \delta \qquad \Rightarrow \delta = 253.5 \text{ s}$$

2) Coconut water naturally contains few spoilage-causing enzymes, which are mainly polyphenol oxidase (PPO) and peroxidase (POD) enzymes. DBD cold plasma has been used to inactivate them. The following table has given both enzymes' residual activity (RA).

CONDITION	TREATMENT TIME (t, min)	$RA = A/A_i$ PPO (%)	POD (%)
Control	0	100	100
1	1	38	45
2	2	12	23

The symbols A and A_i stand for the enzyme activity of the treated sample and initial untreated sample, respectively.

Calculate the average rate constant k (s^{-1}) for both the enzymes and comment on their relative resistance toward cold plasma according to their respective k-values.

Solution: First order kinetic model, $\ln(RA) = -k \cdot t$

For PPO
Condition 1: $\ln(0.38) = -k(60) \qquad \Rightarrow k = 0.016 \text{ s}^{-1}$
Condition 2: $\ln(0.12) = -k(120) \qquad \Rightarrow k = 0.018 \text{ s}^{-1}$

Average k-value for PPO enzyme inactivation $= \dfrac{0.016 + 0.018}{2} =$
0.017 s^{-1}

For POD

Condition 1: $\ln (0.45) = - k (60)$ => $k = 0.013$ s^{-1}
Condition 2: $\ln (0.23) = - k (120)$ => $k = 0.012$ s^{-1}

Average k-value for POD enzyme inactivation $= \dfrac{0.013 + 0.012}{2} =$ 0.0125 s^{-1}

PPO has a faster inactivation rate constant than POD. This clearly shows that the POD enzyme in coconut water is more resistant to cold plasma processing.

3) Imagine a rubber tube is filled with air with an absolute pressure of 700 kPa at 15°C temperature. Calculate what will happen to the pressure inside the tube if the temperature increases by 15°C, assuming there are no leaks and changes in air volume. Further, calculate the number of molecules for a tube having 1.75 L volume at the final condition. Consider ideal gas conditions.

Solution: The change in pressure is occurring only due to the rise in the temperature.

Initial pressure (P_i) of the gas inside the tube $= 7.0 \times 10^5$ Pa
Initial temperature (T_i) of the gas $= 15°C = 288$ K
Final temperature (T_f) of the gas $= 30°C = 303$ K
Final pressure (P_f) of the gas inside the tube $=?$

Applying the ideal gas law in terms of a molecule, $P \cdot V = N \cdot k_B \cdot T$; where P is pressure, V is volume, N is the number of gas molecules, k_B is the Boltzmann constant $(k_B = 1.38 \times 10^{-23}$ J·K$^{-1})$, and T is temperature.

$$P_i \cdot V_i = N \cdot k_B \cdot T_i \quad \text{and} \quad P_f \cdot V_f = N \cdot k_B \cdot T_f$$

$$V_i = \frac{N \cdot k_B \cdot T_i}{P_i} \quad \text{and} \quad V_f = \frac{N \cdot k_B \cdot T_f}{P_f}$$

Given: Initial volume (V_i) of air = Final volume (V_f) of air

$$=> \frac{N \cdot k_B \cdot T_i}{P_i} = \frac{N \cdot k_B \cdot T_f}{P_f} \qquad => \frac{T_i}{P_i} = \frac{T_f}{P_f}$$

$$=> P_f = \frac{T_f \cdot P_i}{T_i} \qquad => P_f = \frac{303 \times (7.0 \times 10^5)}{288}$$

$$=> P_f = 7.36 \times 10^5 \, \text{Pa}$$

For calculating the number of air molecules inside tube having a volume (V_f) of 1.75 L or 1.75×10^{-3} m³, at $P_f = 7.36 \times 105$ Pa and $T_f = 303$ K using the same equation.

$$N = \frac{P_f \cdot V}{k_B T_f}$$

$$\Rightarrow N = \frac{(7.36 \times 10^5) \times (1.75 \times 10^{-3})}{(1.38 \times 10^{-23}) \times 303}$$

$$\Rightarrow N = 3.08 \times 10^{23} \text{ molecules}$$

6.9 Multiple Choice Questions

1. Cold plasma involves _____ ionization _____ having a thermodynamic equilibrium among the electron and heavier species in the plasma.
   a. Complete, with
   b. Complete, without
   c. Incomplete, with
   d. Incomplete, without

2. Cold plasma-induced microbial inactivation happens not due to:

   Reason statements: (1) Chemical reaction with reactive species, (2) Electroporation effect, (3) Dielectric breakdown of the cell and cellular material, (4) UV-based photochemical effect
   a. Statement 1 & 2
   b. Statement 2 & 3
   c. Statement 3 & 4
   d. Statement 4 & 1

3. One of the reasons for the inactivation of food spoilage enzymes by cold plasma can be attributed to the loss of _____ structural properties of the enzyme after exposure.
   a. Quaternary
   b. Tertiary
   c. Secondary
   d. Primary

4. Out of the two plasma generation stages, which stage is associated with: (1) Emission of photons, (2) Formation of radicals, and (3) High energy electron impact?
   a. 1 & 2: Recombination stage; 3: Ionization stage
   b. 2 & 3: Recombination stage; 1: Ionization stage
   c. 3: Recombination stage; 1 & 2: Ionization stage
   d. 1: Recombination stage; 2 & 3: Ionization stage

5. The corona discharge plasma device involves forming discharge volume/area usually on _____ electrode and generates _____ density plasma.
   a. Both, high
   b. Both, low
   c. One, high
   d. One, low

6. The non-thermal plasma possesses the following characteristics: (1) Localized thermodynamic equilibrium, (2) Localized thermodynamic nonequilibrium, (3) Low temperature, (4) Overall charged state. Identify the correct options.
   a. 1 and 4
   b. 2 and 3
   c. 1 and 3
   d. 2 and 4

7. It has been observed that cold plasma treated carbohydrates and proteins have developed better solubility in water, causing the overall food material to uptake water more efficiently and become soft and less firm in texture. This can be attributed to the surface _____ effect of plasma, which can be helpful during soaking and cooking operations.
   a. Etching
   b. Erosion
   c. Scratching
   d. Perforation

8. Cold plasma can degrade pollutants, pesticides, aflatoxins, and food allergens through its _____.
   a. Incomplete ionization pathway
   b. Radiative recombination pathway

   c. Dissociative recombination pathway

   d. Quenched plasma non-recombination pathway

9. Name the chemical augmentation of plasma in which the active species obtained from cold plasma are supplemented with hydrogen from the moisture.

   a. Hydro-plasma

   b. Reduced cold plasma

   c. Plasma-activated water

   d. Plasma-activated moisture

10. Which type of cold plasma device requires a supply of an alternating current with approximately 10 kV ignition voltage, and it is also called the silent discharge? Its dielectric coated electrodes directly comes in contact with the food samples according to its design.

   a. Gliding arc discharge plasma

   b. Dielectric barrier discharge plasma

   c. Microwave powered plasma

   d. Radiofrequency plasma

6.10 Short Answer Type Questions

a. Why is cold plasma called the nonequilibrium plasma? How does the cold plasma possess an overall neutral state without the localized thermodynamic equilibrium?

b. How do thermal and non-thermal plasma different from each other? Explain along with examples.

c. What is the difference between the dissociative and radiative recombination pathways for the recombination stage of the plasma?

d. Explain how cold plasma treatment can affect the nutritional quality of food products and how it affects the physical properties of solid food items?

e. In what manner, the distance of the food sample from the source of cold plasma or its active zone can affect its efficacy?

6.11 Descriptive Questions

a. Describe the steps involved in the formation of cold plasma.

b. What are the microbial and enzymatic inactivation mechanisms by cold plasma processing?

c. With the help of a schematic, explain the principle of two types of cold plasma devices.

d. What are the critical processing factors that influence the efficacy of cold plasma treatment of food products? Elaborate with suitable examples.

e. Point out the various challenges existing in the non-thermal plasma treatment of food products.

6.12 Numerical Problems

1) Effect of DBD cold plasma has been investigated on the total aerobic psychrotrophic bacteria (TAPB) and lactic acid bacteria (LAB) present on fresh fish fillets. The gap between the two electrodes was set at 35 mm, the electric current frequency was fixed at 50 Hz, and a voltage of 80 kV was used for plasma generation. Fish samples were packed in PET plastic tray covered with thick plastic film. These were treated with plasma by keeping it between the electrodes for 2.5–5 min. The results of the microbial analysis have been tabulated below.

CONDITION	TREATMENT TIME (t, min)	Log CFU·g^{-1} TAPB	LAB
Control	0	5.9	4.1
1	2.5	4.8	3.3
2	5	4.2	3.0

Calculate the average decimal reduction time D (min) for both the group of microorganisms and comment on their relative resistance toward cold plasma in terms of D-value.

2) A gas cylinder having a plunger on its top is filled with a specific gas at an absolute pressure of 350 kPa at 10°C. Estimate how the temperature of the gas inside the cylinder changed after pressing the plunger to decrease the volume by 48% and increase the pressure up to 800 kPa. Assume there are no gas leaks and temperature loss to the environment. Further, calculate the number of molecules present in the cylinder having 10 L volume at the final condition. Consider ideal gas conditions.

References

Devi, Y., Thirumdas, R., Sarangapani, C., Deshmukh, R. R., & Annapure, U. S. (2017). Influence of cold plasma on fungal growth and aflatoxins production on groundnuts. *Food Control*, 77, 187–191.

Suggested Readings

Ekezie, F. G. C., Sun, D. W., & Cheng, J. H. (2017). A review on recent advances in cold plasma technology for the food industry: Current applications and future trends. *Trends in Food Science & Technology*, 69, 46–58.

Misra, N. N., Schlüter, O., & Cullen, P. J. (Eds.). (2016). *Cold Plasma in Food and Agriculture: Fundamentals and Applications*. Academic Press, San Diego, USA.

Niemira, B. A. (2012). Cold plasma decontamination of foods. *Annual Review of Food Science and Technology*, 3, 125–142.

Niemira, B. A., & Gutsol, A. (2011). Non-thermal Plasma As a Novel Food Processing Technology. In H. Q. Zhang, G. V. Barbosa-Canovas, V. M. Balasubramaniam, C. P. Dunne, D. F. Farkas, & J. T. C. Yuan (Eds.), *Nonthermal Processing Technologies for Food*. Wiley-Blackwell and IFT Press, Chichester, West Sussex, UK. pp. 271–288.

Answers for MCQs (sec. 6.9)

1	2	3	4	5	6	7	8	9	10
d	b	c	a	d	b	a	c	c	b

7

OZONE PROCESSING

7.1 Principles

Ozone (trioxygen, O_3) gas is formed naturally by converting the localized atmospheric oxygen during lighting. It exists naturally in our earth's atmosphere (**Fig. 7.1**).

The ozone gas has high reactivity, good penetrability, and it decomposes spontaneously into numerous free radicals and non-toxic oxygen and leaves no residue. Artificial ozone can be generated through corona discharge, photochemical, radiochemical, and electrolytic methods. UV-based photochemical methods and corona discharge methods are prevalent in ozone production. In the corona discharge method, the electrodes are kept apart at a certain distance with the help of some dielectric material (**Fig. 7.2**). This process strictly requires pure dry air or oxygen gas as a feed, free of dust, oil, or any foreign compound or material. Under the influence of the electric field, the oxygen molecules are made to split into oxygen atoms, which is highly unstable and capable of forming other active oxygen radicals. The oxygen atoms, when combined with other oxygen molecules (O_2), generate ozone (O_3) (Eq. 7.1). The resulting gas mixture contains about 1–3% of ozone if dry air has been used, or about 3–6% ozone is formed if pure oxygen gas was used as the feed gas.

$$O_2 + e^- \rightarrow 2O$$
$$2O + O_2 \rightarrow 2O_3 \tag{7.1}$$

Similar ozone production can also be achieved through the photochemical action of UV (**Fig. 7.3**). The wavelength of 185 nm is mainly responsible for ozone generation. It does not necessarily require dry air, and the overall system is relatively cheaper than other methods. However, the percentage concentration is around

DOI: 10.1201/9781003199809-7

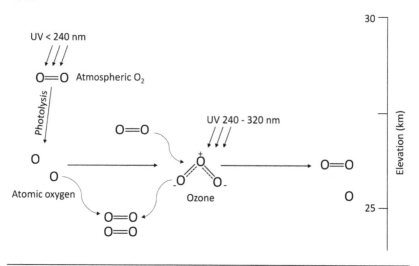

Figure 7.1 Formation and destruction of ozone in the earth's stratosphere. (Adapted from Chawla et al. (2012) with minor changes)

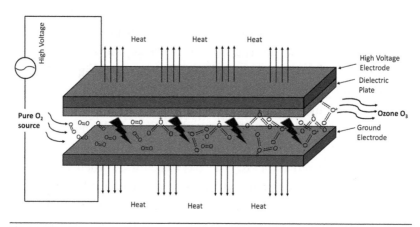

Figure 7.2 Ozone production by corona discharge method.

0.1% ozone when dry air was used as the feed gas, lower than the corona discharge method. In addition to the UV-based photochemical and electric field-based corona discharge methods, ozone can be produced through an electrolytic method (**Fig. 7.3**). It follows an electrochemical pathway to split water molecules into hydrogen and oxygen atoms through electrolysis. During the electrolytic decomposition of water, hydrogen molecules are separated from the water-gas mixture, allowing the oxygen atoms to produce ozone and diatomic oxygen.

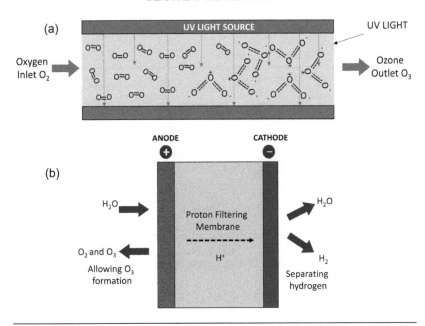

Figure 7.3 Ozone production by (a) UV light and (b) electrolysis method.

7.2 Mechanisms of Quality Changes in Food

Ozone has a molecular mass of 48 $g \cdot mol^{-1}$, a density of 2.14 $kg \cdot m^{-3}$, and a boiling point of $-112°C$. It is soluble in water. Ozone is relatively more stable in gaseous form than aqueous form. Ozone in the aqueous phase is commonly used in water purification and disinfection. Molecular ozone (O_3) and its decomposition derivatives, including ions, radicals, and intermediates, such as superoxide radical ion ($^{\bullet}O_2^{-}$), hydroperoxide radical ($^{\bullet}HO_2$), and hydroxyl radical ($^{\bullet}OH$), etc., can initiate different chemical reactions that are mainly responsible for the antimicrobial property and pesticide degradation ability of the ozone. Unlike other non-thermal technologies, ozone is very effective against a broad spectrum of microorganisms, including bacteria, bacterial spore, yeast, fungi, fungal spore, virus, and protozoa. Free radicals generated during the ozonation process are very unstable and easily oxidize the vital components of the microbial cell. First, the molecular ozone and derived free radicals chemically attack the cell wall or membrane components by oxidizing the polyunsaturated fatty acids, glycoproteins, membrane-bound enzymes, and glycolipids resulting in the membrane or cell wall breakage. This causes leakage

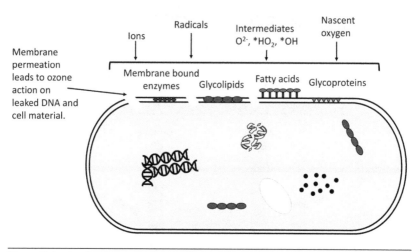

Figure 7.4 Inactivation mechanism of microorganism due to ozone processing.

of cellular and genetic (DNA) materials, which is again exposed to the highly oxidative surrounding of ozone (**Fig. 7.4**).

Ozone also degrades spore coats and virus capsids as it can degrade the proteins, peptidoglycan, and associated sulfhydryl groups. Following similar chemical pathways, ozone can deteriorate pesticides, mycotoxins, and unpleasant odor-causing volatiles generated from waste material or physical food processing. Even though ozone can potentially denature protein materials associated with food spoilage enzymes, little research on enzyme inactivation is available in the literature. Ozone treatment may or may not affect food material's nutritional and sensory properties depending on the target food sample and processing conditions. A more extended exposure period and high ozone concentrations can deteriorate (via prolonged oxidization) nutrients, such as antioxidants, vitamins, polyphenols, and sensory attributes (mainly flavor and aroma).

7.3 Equipment and Its Operation

Ozone treatment can be conducted using either the gaseous or aqueous phase of ozone for food processing. In general, the essential components of the ozone-producing system (**Fig. 7.5**) consist of the inlet feed gas (pure oxygen or air), ozone generator, source of electric power, contactor (only to obtain ozone in water phase), reactor vessel, destruction unit for surplus gas, and analyzer for measuring

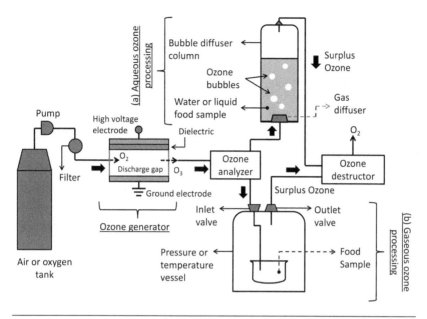

Figure 7.5 Schematic of (a) aqueous and (b) gaseous ozone processing units.

ozone concentration. During ozone generation, the temperature must be kept low, as the temperature above 35°C causes the ozone gas to break down into oxygen ions and associated molecules.

The corona discharge type generators need dry air or a pure form of oxygen, and if the air is used, its moisture is separated by condensing the water vapors at −65°C before feeding to the system. This is done majorly to improve ozone treatment effectiveness and avoid the formation of nitrogen oxides that promote corrosion of electrode and metal surface. During ozone production, the corona discharge system usually operates at 50–60 Hz frequency and > 20 kV voltage and the advanced systems can also operate at 1–2 kHz with a 10-kV voltage supply. According to the target application of ozone treatment, there can be two types of devices to mix the ozone properly in water; bubble diffuser chamber and turbine agitated reactor. The surplus ozone is either destroyed by catalytic decomposition, diluted with air before releasing, or absorbed in moist granular activated carbon for safety purposes.

Many researchers have performed storage studies in the presence of ozone gas. It comes under modified atmospheric storage conditions, where primarily ozone gas is mixed with another gas like dry air and continuously supplied to the storage area at a specific pressure

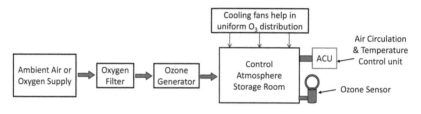

Figure 7.6 Block diagram of food storage room showing ozone application under controlled atmospheric conditions.

(**Fig. 7.6**). Continuous monitoring of ozone gas is necessary to achieve the desired ozone level in the room. The room must have proper airflow of cooling air to avoid localized accumulation or build-up of ozone gas in specific room areas.

7.4 Critical Processing Factors

7.4.1 Type of Product

The efficacy of the ozone treatment varies for solid and liquid food materials having different properties. Ozone works better on the surface and cannot penetrate deeper into the product, primarily solids. Maintaining a pH below 7 is necessary during aqueous ozone treatment to slowdown ozone decay as it reacts with the water. In this sense, if the pH of the liquid food product is below 7, better inactivation of microorganisms might be visible during treatment. Product temperature and moisture content also affect the efficacy of the treatment. High product temperature can diminish the effect of ozone as it may speed up the decay of ozone. The presence of specific organic components may interfere with the action of ozone by competing for O_3 and microorganisms.

7.4.2 Target Microorganism or Enzyme

Every material or target microorganism or target enzyme has a specific ozone demand. Generally, bacteria in the exponential growth phase are more sensitive to any harsh processing environment than the stationary phase. A large initial microbial count and high enzyme concentration can decrease the efficacy of the treatment. After getting neutralized by surrounding unwanted organic compounds in the sample, the remaining or residual ozone is primarily responsible for the antimicrobial action of ozone. Even though ozone can inactivate many microorganisms, spores are usually more resistant than vegetative cells.

The effect of ozone on spoilage enzymes is not much explored in the literature. First-order kinetics, Weibull distribution model (refer Eq. 3.8), and Chick-Watson model (Eq. 7.2) have been commonly used to predict the inactivation trends primarily for microorganisms. The Chick-Watson model is similar to a first-order kinetic model with an additional term of dissolved ozone concentration (C) in the sample.

$$\ln\left(\frac{N}{N_i}\right) = k \cdot C \cdot t \tag{7.2}$$

7.4.3 Equipment and Mode of Operation

The primary process variables affecting the effectiveness of ozone processing are the use of gaseous or aqueous phase of ozone, exposure time, and concentration of ozone or concentration of residual available for further reactions. The amount of ozone and the reactive species generated during ozone formation primarily depends on the gas (dry air or pure oxygen) and means of production (such as corona discharge or UV excitation). High ozone concentration, larger surface area for exposure, and longer contact time with ozone can enhance the microbial inactivation and degradation of unwanted pesticides and mycotoxins process and may compromise the food product's nutritional and sensory qualities.

7.5 Practical Example

We will consider one ozone treatment-based research work as a practical example. The study involves ozone treatment of blueberry fruit, separately inoculated with *Escherichia coli* 0157:H7 and *Salmonella enterica* (Bialka et al., 2007). The effect of ozone in both gaseous and aqueous phases has been tested on the microorganisms. *E. coli* 0157:H7 and *S. enterica* are both pathogenic Gram-negative bacteria. In this study, bacterial counts were quantified as colony forming unit (CFU) per g of the sample taken to make the inoculum for the microbial growth medium (plated agar).

Ozone treatment condition: The ozone generator produced gas with 5% (w/w) ozone at a flow rate of $0.34 \text{ m}^3 \cdot \text{h}^{-1}$. A total of 18 fruits were treated for up to 64 min by subjecting them to; (a) continuous flow of ozone gas, (b) ozone gas pressurized at 83 kPa in a closed treatment chamber, and (b) aqueous ozone with up to 21 mg/L ozone concentration at 4°C. During processing with aqueous ozone, fruits were dipped in a flask containing 500 mL water, and ozone gas was sparged into it. For untreated control samples in the case of aqueous ozone treatment, the air was sparged instead of ozone. The effect of ozone processing has also been investigated on color and sensory properties. The blueberry fruits were inoculated with approximately 10^6 CFU/g (initial count).

Observations after ozone treatment: A more significant impact of ozone exposure time was observed on both the microorganisms. In the case of gaseous ozone treatment for 64 min, a maximum reduction of about 3 log cycles for *S. enetrica* was observed after pressurized ozone treatment. Furthermore, a maximum reduction of about 2.2 log cycles for *E. coli* was detected after continuous ozone treatment. On the other hand, dipping the inoculated fruits in an aqueous ozone solution (21 mg/L) for 64 min inactivated 5.2 and 6.2 log counts of *E. coli* and *S. enterica*, respectively. The ozone-treated blueberries did not significantly change their color parameters and sensory properties.

7.6 Challenges

Ozone processing is gaining more attention as an eco-friendly non-thermal technology due to its excellent food processing potential. Ozone is very effective against a broad spectrum of microorganisms and can degrade pesticides and mycotoxins. It is a cost-effective non-thermal technique. The most advantageous factor of ozone treatment is that it does not leave any residue after processing compared to chemical-based disinfection methods related to chemicals, such as chlorine, ethyl alcohol, formaldehyde, etc. Ozone can also be used as a fumigation gas, which is better than other conventional chemicals, such as SO_2. It has been used for various food materials such as meat, poultry, fish, spices, fruits, vegetables, beverages, dairy, etc. This technology is also applicable for treating both frozen and fresh food items. Ozone works as an excellent surface disinfectant for food contact surfaces. It does not affect most of the quality attributes of food products, including its texture, appearance, and nutritional properties. However, treatment with high ozone concentrations and longer exposure time can deteriorate the vitamins, polyphenols, color, and affect sensory properties related to food flavor and aroma. The effect of ozone treatment varies for different microorganisms, their physiological condition, type of food product, temperature, pH, presence of any organic material capable of absorbing or neutralizing the ozone and its secondary active compounds, the physical state of ozone, etc. The strong dependence of ozone treatment on numerous factors makes it difficult to determine its compatibility with the sample and target microorganism or enzyme and the most suitable processing conditions. The ozone must be produced on site of its usage because of its unstable nature. Rapid decompositions of ozone in the water phase (ozone activated water) limit the antimicrobial effect of ozone only up to the food surface, which allows the microorganisms embedded in the food product to survive. The strong oxidizing power of ozone leads to corrosion of the surface of the metal food containers. Ozone is harmful and toxic for humans because of its high oxidizing capabilities and spontaneous formation of free radicals. Inhaling ozone can affect our lungs by causing diseases, such as respiratory tract inflammation, capillary hemorrhage, pulmonary edema, various lung diseases, and exposure to a heavy concentration of ozone (> 500 ppm) may even induce mutagenic defects and death in the long run. A proper system for eliminating or absorbing the excess ozone must be installed, and adequate safety measures for the early identification and prevention of any possible leakage.

7.7 Summary

- Reactive behavior and spontaneous decomposition into numerous free radicals and nontoxic oxygen leaving no residue makes ozone treatment a safe non-thermal technology.
- Artificial ozone can be generated through corona discharge, photochemical (UV), radiochemical, and electrolytic methods.
- During the production of ozone with corona discharge, 1–3% and 3–6% of ozone is produced when dry air and pure oxygen gas are used as the feed gas, respectively.

- The photochemical action of UV (wavelength of 185 nm) is mainly responsible for ozone generation, but the yield is poor than the corona discharge method.
- Decomposition derivatives of ozone, including ions, radicals, and intermediates, such as superoxide radical ion ($^{\bullet}O_2^-$), hydroperoxide radical ($^{\bullet}HO_2$), and hydroxyl radical ($^{\bullet}OH$), which are responsible for its antimicrobial property.
- Ozone oxidizes cell membrane, membrane-based proteins, glycolipids, and causes leakage of cell organelles and allowing further damage to the exposed DNA.
- Ozone is very effective against a broad spectrum of microorganisms and can degrade pesticides, mycotoxins, and unpleasant odor-causing volatiles from waste material.
- Temperature above 35°C can breakdown ozone into oxygen ions and associated molecules and diminish its oxidizing power.
- During aqueous ozone treatment, maintaining a pH below 7 is necessary to slowdown the decay of ozone as it reacts with the water.
- The strong oxidizing power of ozone makes it harmful to humans, and also allows it to erode the surface of the metallic food containers.

7.8 Solved Numerical

1) Ozone treatments were found to be very effective against oocysts of *Cryptosporidium pavum*. The details about the inactivation of the oocysts (suspended in a buffer) by applying aqueous ozone bubbled into the solution using a diffuser.

INACTIVATION ($-\log (N/N_i)$)	OZONE CONCENTRATION (C, mg·L^{-1})	EXPOSURE TIME (t, min)
−0.5	0.36	2.97
−2	0.86	3.58

Calculate the average inactivation rate constant (k) and predict the inactivation ($\log N/N_i$) at $t = 3.47$ min, and concentration is 81% of the final or maximum ozone concentration. [*C. pavum* is a parasitic protozoan that affects domestic and wild animals. The oocysts are a specific stage of the life cycle of such a protozoan, which has a hard and thick wall].

Solution:

$$\ln(N/N_i) = -k \cdot C \cdot t \quad \Rightarrow 2.303 \times \log(N/N_i) = -k \cdot C \cdot t$$

a. $2.303 \times (-0.5) = -k \times 0.36 \times 2.97 \quad \Rightarrow k = \frac{2.303 \times 0.5}{0.36 \times 2.97} = 1.08 \text{ L} \cdot \text{mg}^{-1} \cdot \text{min}^{-1}$

b. $2.303 \times (-2) = -k \times 0.86 \times 3.58 \quad \Rightarrow k = \frac{2.303 \times 2}{0.86 \times 3.58} = 1.49 \text{ L} \cdot \text{mg}^{-1} \cdot \text{min}^{-1}$

Average inactivation rate constant of the oocyst, $k_{avg} = \dfrac{1.08 + 1.49}{2} = 1.29 \text{ L} \cdot \text{mg}^{-1} \cdot \text{min}^{-1}$

Now predicting the inactivation value at $t = 3.47$ min and ozone concentration $(C) = 0.81 \times 0.86 = 0.70 \text{ mg} \cdot \text{L}^{-1}$ by using derived first order kinetics model.

$$2.303 \times \log(N/N_i) = -1.28 \times 0.70 \times 3.47$$

$$\Rightarrow \log(N/N_i) = \frac{-1.28 \times 0.70 \times 3.47}{2.303} \quad \Rightarrow \log(N/N_i) = 1.35$$

2) Orange juice has been treated with gaseous ozone, which was released into the sample using a bubble diffusion system. Loss of vitamin C (ascorbic acid) was measured for different exposure times and different ozone concentrations (OC).

S.NO.	TIME (t, min)	RESIDUAL VITAMIN-C CONCENTRATION (C/C_i) AT THREE DIFFERENT OZONE CONCENTRATIONS		
		0.6% OC	1.8% OC	3% OC
1	2	1	1	1
2	4	0.78	0.70	0.60
3	10	0.53	0.42	0.38

Determine the value of k_z that stands for the change in degradation rate constant of vitamin-C (k) concerning the change in ozone concentration (OC). Consider zero-order kinetics for estimating both k and k_z.

Solution: Zero-order kinetic model equation, $C = C_i - k \cdot t$

Where C is the concentration of vitamin-C after treating for time t, C_i is the initial concentration of vitamin-C of the untreated juice, and k is the degradation rate constant whose negative sign represents that

C decreases with treatment time. This equation is similar to the linear equation of a straight line ($y = m \cdot x + c$), in which m is the slope of the line. In our case, k is the slope.

Overall slope $= \frac{y_2 - y_1}{x_2 - x_1}$

1. $-k_1 = \frac{0.53 - 1}{10 - 0} = 0.047$ min^{-1} {at 0.6% OC}

2. $-k_2 = \frac{0.42 - 1}{10 - 0} = 0.058$ min^{-1} {at 1.8% OC}

3. $-k_3 = \frac{0.28 - 1}{10 - 0} = 0.072$ min^{-1} {at 3% OC}

Again, using zero-order for calculating k_z. $\qquad k = k_i - k_z \cdot t$

Similarly calculating slope (k_z);

$$k_z = \frac{k_3 - k_1}{OC_3 - OC_1} = \frac{0.072 - 0.047}{3 - 0.6}$$

$$k_z = 0.01 \text{ min}^{-1} \cdot (OC\%)^{-1}$$

3) A study related to the solubility of ozone in water was conducted. The ozone concentration in dissolved form (C_L, ozone in liquid phase) and gaseous form (C_g) has been noted down at specific conditions of 20°C, at atmospheric pressure, and for a 150-mL total volume of water. Ozone gas was bubbled into the sample at 10 L·h^{-1}, having a pH of 5.5.

a. Calculate the solubility ratio (R_l) and comment on its trend.

S. NO.	TIME (t, min)	C_G (mg [O_3]· (L Water)$^{-1}$)	C_L (mg [O_3]· (L Water)$^{-1}$)
1	2.5	10	5.0
2	5	44	14.08
3	10	47	14.57
4	15	48	14.40

b. The solubilized ozone is not stable in water, so after reaching a certain equilibrium, it decreases by either going back to the gaseous phase or degrading into O_2. Calculate the degradation rate as per first-order kinetics and estimate the half-life of ozone in water.

S. NO.	TIME AFTER EQUILIBRIUM (t, min)	In (C_L/C_{Li})
1	10	−1.2
2	130	−2.5

C_{Li} is the initial concentration of the dissolved O_3 in water.

Solution:

a. Solubility ratio, $R_t = \dfrac{C_L}{C_g}$

(1) $R_t = \dfrac{5}{10} = 0.5$, (2) $R_t = \dfrac{14.08}{44} = 0.32$, (3) $R_t = \dfrac{14.57}{47} = 0.31$,

(4) $R_t = \dfrac{14.4}{48} = 0.30$

Initially, the solubility ratio (R_t) of ozone decreased with time. Then it reached an equilibrium concentration (steady-state), where the conversion of O_3 from liquid to gaseous phase slows down.

b. Now estimating O_3 degradation rate (k) as per the first-order kinetics;

$$\ln\left(\frac{C_L}{C_{Li}}\right) = -k \cdot t \qquad => k = -\frac{\ln\left(C_L/C_{Li}\right)}{t}$$

Applying equation of slope of straight line to solve for k;

$$slope = \frac{y_2 - y_1}{x_2 - x_1} \qquad => k = \frac{-(-2.5 - (-1.2))}{130 - 10}$$

$$=> k = 0.011 \text{ min}^{-1}$$

Finally, calculating the half-life ($t_{1/2}$) of ozone in water represents the time required when the dissolved O_3 becomes half of its initial equilibrium point. $C_L = 0.5 \times C_{Li}$

$$t_{1/2} = -\frac{\ln\left((0.5 \times C_{Li})/C_{Li}\right)}{k} \qquad => t_{1/2} = \frac{0.693}{k}$$

$$=> t_{1/2} = \frac{0.693}{0.011} \qquad => t_{1/2} = 63 \text{ min}$$

7.9 Multiple Choice Questions

1. In corona discharge, the ozone is generated by the action of _____, and in the case of UV-based technology, ozone is produced due to _____ action.
   a. Photochemical, electric field
   b. Electric field, photochemical
   c. Electron impact, photothermal
   d. Photothermal, electron impact

2. Select the true statements.
   (1) More ozone is generated when dry air is used rather than oxygen as the feed gas. (2) Moisture-free air is not compulsory for UV-based ozone generators. (3) Higher ozone concentration is produced through the corona discharge method. (4) Corona discharge ozone generator is cheaper than UV-based ozone generator.
   a. Statement 4 & 1
   b. Statement 3 & 4
   c. Statement 2 & 3
   d. Statement 1 & 2

3. Chemical reactions with molecular ozone and its secondary free radicals can potentially inactivate: (1) Bacteria, (2) Yeast and fungi, (3) Bacterial and fungal spores, (4) Virus, and (5) Protozoan.
   a. 1, 2, 3, 4, and 5
   b. 1, 2, 4, and 5
   c. 1, 4, and 5
   d. 1 and 2

4. The respective operating conditions of corona discharge and UV-based ozone generators are:
   a. 20–30 Hz frequency with > 20 kV voltage and 254 nm wavelength
   b. 50–60 Hz frequency with 10–20 kV voltage and 185 nm wavelength
   c. 20–30 Hz frequency with 10–20 kV voltage and 254 nm wavelength
   d. 50–60 Hz frequency with > 20 kV voltage and 185 nm wavelength

5. Ozone is more stable in the _____ phase. In the aqueous phase, a pH of _____ and a temperature of _____ are necessary to delay ozone decay and improve its effectiveness.
   a. Gaseous, below 7, below 35°C
   b. Aqueous, above 7, below 35°C
   c. Gaseous, below 7, above 35°C
   d. Both gaseous and aqueous, above 7, above 35°C

6. Two statements are given to compare the solubility of gaseous ozone in water with a few other gases. (1) Ozone is more soluble in water than nitrogen and oxygen, (2) Chlorine and carbon dioxide are less soluble in water than ozone.
   a. Statement 1 is false, and statement 2 is true
   b. Statement 1 is true, and statement 2 is false
   c. Both statements 1 and 2 are true
   d. Both statements 1 and 2 are false

7. Ozone production using corona discharge system requires _____.
   a. Strictly pure dry air or oxygen
   b. Strictly pure moist air or oxygen
   c. Strictly pure dry air or moist oxygen
   d. Strictly pure moist air or dry oxygen

8. Before feeding air to the corona discharge system, its moisture is separated by condensing the water vapors at −65°C to _____.
   a. Decrease operational voltage and avoid sparking
   b. Improve feed gas usage and avoid the formation of unwanted gases
   c. Decrease the operational frequency and avoid O_3 production delay
   d. Improve ozone treatment effectiveness and avoid electrode corrosion

9. Two types of contactors widely used to mix the ozone properly in water are:
   a. Ozone dissolver and baffled circulating tube
   b. Ozone dissolver and turbine agitated reactor
   c. Bubble diffuser chamber and turbine agitated reactor
   d. Bubble diffuser chamber and baffled circulating tube

10. Ozone in the water phase (ozone activated water) _____ the antimicrobial effect of ozone, which _____ the microorganisms embedded in the food product to survive.
    a. Improves, does not allow
    b. Limits, allow
    c. does not change, may or may not allow
    d. Changes, do not allow

7.10 Short Answer Type Questions

a. Differentiate natural and artificial ozone and explain why ozone processing is considered safe and nontoxic to food? State the general components of the ozone production system.
b. How is corona discharge different from UV-based photochemical systems and electrolytic decomposition of a water-based system for ozone production?
c. Explain the Chick-Watson model used for microbial inactivation kinetics.
d. How different types of food products can affect the efficacy of ozone processing?
e. Highlight the advantages of ozone processing.

7.11 Descriptive Questions

a. Describe the microbial inactivation mechanism of antimicrobial action of ozone.
b. With a schematic, discuss the method of gaseous ozone production through the corona discharge system.
c. How can different target enzymes or microorganisms and equipment and modes of operation affect the efficacy of ozone processing?
d. Discuss the working principle of ozone-activated water with a diagram.
e. Discuss the challenges and disadvantages in processing food products with ozone.

7.12 Numerical Problems

1) Quantification of the solubility of gaseous ozone in water (concentration of dissolved Ozone, C) at different partial pressure (P, kPa) of

ozone on water and different pH of water has been studied separately. Calculate Henry's law constant of solubility (k_H, kPa·L·mol^{-1}) and comment on the effect of changing P or pH on k_H. Water was kept at 10°C, ionic strength of the water was maintained at 0.15 M with sodium phosphate. During variable partial pressure of ozone, pH was fixed at 7, and during variable pH, partial pressure of ozone was fixed at 2 kPa. The experimental data have been tabulated below. [HINT: Henry's Law; $P = k_H \times C$]

S. NO.	P (kPa)	$C = [O_3] \times 10^4$ mol·L^{-1} (MOLES OF O$_3$ PER LITER OF WATER)
1	0.5	0.9
2	2.0	3.2
3	3.5	5.0

S. NO.	pH	$C = [O_3] \times 10^4$ mol·L^{-1} (MOLES OF O$_3$ PER LITER OF WATER)
1	9	1.9
2	7	3.5
3	2.5	5.7

2) Efficacy of ozonated water for inactivating *Shigella sonnei* (Gram-negative pathogen) inoculated on shredded lettuce was researched. An initial count of the microorganism was maintained at 10^8 CFU·g^{-1}. The food samples were treated with ozone-activated water for 60 s. The logarithmic survival fraction (log S = log N/N_0) for three ozone concentrations (C) in water was noted. For ozone concentrations of 0.5, 1.6, and 2.2 ppm in water, log S values were −0.3, −3.7, and −5.6. Apply Chick-Watson's kinetic model to estimate the inactivation rate (k, L·mg^{-1}·s^{-1}). Also, predict the survival fraction of *S. sonnei* if the sample is treated with 1.2 ppm ozone concentration dissolved in water and for a treatment time of 60 s. [HINT: Solve the kinetic model considering $C \times t$ as x-axis or independent variable.]

References

Bialka, K. L., & Demirci, A. (2007). Decontamination of *Escherichia coli* O157: H7 and *Salmonella enterica* on blueberries using ozone and pulsed UV-light. *Journal of Food Science, 72*(9), M391–M396.

Suggested Readings

Brodowska, A. J., Nowak, A., & Śmigielski, K. (2018). Ozone in the food industry: Principles of ozone treatment, mechanisms of action, and applications: An overview. *Critical Reviews in Food Science and Nutrition*, *58*(13), 2176–2201.

Cullen, P. J., & Tiwari, B. K. (2012). Applications of Ozone in Fruit Processing. In S. Rodrigues & F. A. N. Fernandes (Eds.), *Advances in Food Processing Technologies*. CRC Press, Taylor and Francis Group, Boca Raton, Florida, USA. pp. 185–202.

O'Donnell, C., Tiwari, B. K., Cullen, P. J., & Rice, R. G. (Eds.). (2012). *Ozone in Food Processing*. Wiley-Blackwell, Chichester, West Sussex, UK.

Zhang, H. Q., Barbosa-Cánovas, G. V., Balasubramaniam, V. B., Dunne, C. P., Farkas, D. F., & Yuan, J. T. (Eds.). (2011). *Non-thermal Processing Technologies for Food*. Wiley-Blackwell, Chichester, West Sussex, UK.

Answers for MCQs (sec. 7.9)

1	2	3	4	5	6	7	8	9	10
b	c	a	d	a	b	a	d	c	b

8

IRRADIATION

8.1 Principle

Radiation is energy emitted by a source that travels through a particular space or medium. Different energy waves produced by various sources can be distinguished on an electromagnetic spectrum and further categorized based on wavelength and frequency (**Fig. 8.1**).

Irradiation is the process of exposing any material to radiation. When radiation possesses high energy or high frequency, it can be considered as ionizing radiation such as alpha particles (identical to helium ions with two electrons missing, He^{2+}), beta (β) rays, or electron beams, X-rays, and gamma (γ) rays. Ionizing radiations have high-energy electromagnetic waves that produce ions or charged particles in the materials whose molecules (M) receive these radiations. For instance, getting exposed to a photon of energy ($E = h \cdot \nu$) removes a bound electron (e^-) from the molecule (M), leaving a positively charged molecule (Eq. 8.1). Different kinds of lights, gamma irradiation, and X-rays are part of the electromagnetic spectrum and demonstrate dual nature as per quantum theory. They can be considered as waves or energy packets called photons. Photons are electromagnetic radiation with zero rest mass, zero charges, and move at the speed of light.

$$M \xrightarrow{h\nu} M^+ + e^- \tag{8.1}$$

Where h is the Plank's constant equal to 6.626×10^{-34} $m^2 \cdot kg \cdot s^{-1}$ and ν (Hz) is the frequency, which can be estimated as the ratio of the speed of light ($c = 3 \times 10^8$ $m \cdot s^{-1}$) and wavelength (λ) (Eq. 8.2). Greater wave frequency corresponds to the radiation's higher energy and penetration ability.

$$\nu = \frac{c}{\lambda} \tag{8.2}$$

Photons can interact with any matter in three ways; *Compton effect*, *pair production*, and the *photoelectric effect*. These interactions are

DOI: 10.1201/9781003199809-8   **143**

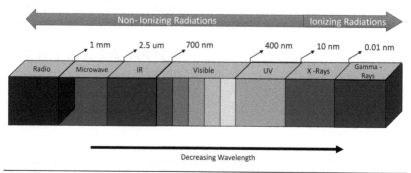

Figure 8.1 Electromagnetic wave spectrum.

greatly influenced by the type of matter and the photon energy (E). In the *Compton effect*, the photon considered a particle collides with an orbital electron through a certain angle. It gets deflected after passing some of its energy onto it, causing the electron to eject (recoil electron) with certain derived kinetic energy (**Fig. 8.2**). It is the most dominant mechanism of the interaction of photons and any tissue-type material. The *pair production* occurs when high energy photons (such as high energy X-rays and gamma rays) pass near the nucleus of an atom

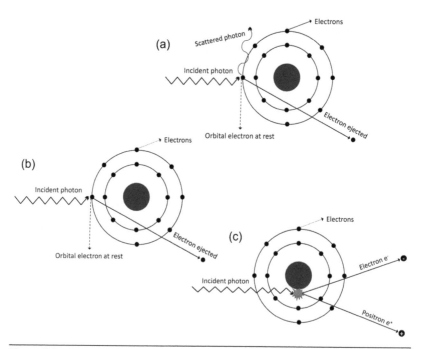

Figure 8.2 Interaction mechanisms between particles of photon and matter; (a) *Compton effect*, (b) *Photoelectric effect*, and (c) *Pair production*.

may experience strong field effects from the nucleus and causing the photon to convert into a positron (e^+) and an electron (e^-). During the *photoelectric effect*, the incident photon is completely absorbed into the atom resulting in the disappearance of the photon and ejection of an orbital electron with kinetic energy derived from the photon's energy. On the other hand, electromagnetic waves having relatively low energy, such as radio waves, microwave, and infrared, cannot produce ions or charged particles in exposed materials and are thus considered nonionizing radiations. This chapter focuses on ionizing radiation, which can break chemical bonds and ionize atoms. The major objectives of irradiation include the killing of insects, pests, spoilage, and pathogenic microbes, along with the inhibition of sprouting in certain food products.

Ionizing radiations primarily have wavelength less than or equal to 200 nm. However, 'electron beam', 'gamma irradiation', and 'X-rays' are majorly used in food preservation. The shorter the radiation wavelength, the higher is the penetration power and its lethal effect. Since they ionize molecules in their propagation path without an appreciable temperature rise, it is also known as cold sterilization.

Agricultural products and prepacked food products can be subjected to controlled irradiation to induce radiolysis (the process of molecule dissociation due to ionizing radiation) in important biomolecules and attack on insects, parasites, and pathogens. Depending on the desired effect, different doses of irradiation are preferred. The energy gained by an electron traveling through a voltage difference of one volt is defined as one eV (1 eV = 1.6×10^{-19} J). Million or mega electron volt (MeV) is majorly used to describe the intensity of irradiation or energy carried by radiation. Subsequently, the 'absorbed radiant energy by an object' or the 'dose of irradiation' is measured in gray (Gy). One unit of gray (Gy) refers to 1 Joule radiant energy absorbed per kilogram of the material (food). Other units such as rad (1 Gy = 100 rad) is also used to quantify the ionizing radiation dose. Natural radioactive isotopes can generate gamma rays. Beta rays and X-rays can be produced from natural radioactive isotopes and electrical machines. Isotopes are the members of the family of an element having the same number of protons but a different number of neutrons in their nucleus, which means elements that share the same atomic number but different atomic mass due to differences in the number of neutrons in their nuclei.

Various irradiations in their increasing frequency or wavelength can be arranged in electron beam or beta rays, X-rays, and gamma rays.

8.2 Mechanism of Lethality

8.2.1 Microbial Inactivation

The effect of ionizing radiation on microbial inactivation is significant, and it highly depends on the susceptibility of the pathogen or the spoilage bacteria to radiation. It is worthwhile to note that there is a range of diversity in decimal reduction (D-values: the dose required for 10-fold reduction) of the bacterial population to irradiation treatment, with variability observed due to matrices and respective applications. For example, *Cronobacter* spp. showed a decimal reduction dose of 4.83 kGy in dehydrated milk and a value of 0.76 kGy in milk powder, both at room temperature. Irradiation can, however, not be an effective treatment for the reduction of microbial toxins. *Staphylococcal* toxins may have a value as high as 95 kGy as the decimal reduction dose. Fungal toxins, such as aflatoxins and ochratoxins, are degradable by gamma irradiation. The effects of irradiation on microbial inactivation can be divided into *direct* and *indirect* effects (**Fig. 8.3**). The *direct* effect involves the radiation directly attacking the genetic material and breaking the bonds between the base pairs, resulting in cell inactivation. In the *indirect* method, the water in food undergoes radiolysis to form

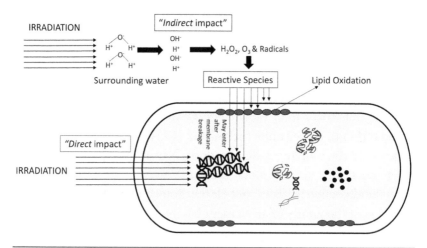

Figure 8.3 Summary of the mechanism of microbial inactivation by irradiation (ionizing radiation).

hydrogen ions and hydroxyl radicals, which irreversibly attack the genetic materials, cell organelles, and cell membrane. On exposure to irradiation, water gets ionized to a positively charged molecule and an electron is produced. The positive water ion dissociates into hydroxyl radical and an H^+ ion. The radicals have unpaired electrons forming hydrogen peroxides, hydrogen gas, and water (Eq. 8.3).

$$H_2O \rightarrow H_2O^+ + e^-$$
$$H_2O^+ \rightarrow OH^\bullet + H^+$$
$$e^- + H_2O \rightarrow OH^- + H^\bullet$$
$$2H^\bullet \rightarrow H_2 \qquad (8.3)$$
$$2(OH)^\bullet \rightarrow H_2O_2$$
$$H^\bullet + OH^\bullet \rightarrow H_2O$$

In the presence of hydroperoxyl radical ($\bullet OOH$) and superoxide radical ($\bullet O_2^-$), more hydrogen peroxide is formed. In addition to H_2O_2, ozone (O_3) can also be produced from oxygen while irradiation, leading to greater lethality due to their oxidizing action. The decimal reduction values are decreased due to the increased effectiveness of the process.

8.2.2 Spoilage Enzymes Inactivation

Enzymes are the most resistant to irradiation, making them highly important in food preservation. The *direct* and *indirect* mechanisms as discussed earlier also have an impact on enzymes. Irradiation can activate the substrate or unfold the enzyme molecule, making the site of attacks more accessible. This can lead to an increased reaction rate, accelerating spoilage. On the contrary, destruction of the 3-dimensional tertiary structure of the enzyme contributes to the inactivation of enzymes, such as lipase. Sometimes food enzymes need high doses (50 kGy) for inactivation. When enzymes are irradiated, the radiolytic products formed react with the amino acids of the enzymes, majorly histidine, tyrosine, tryptophan, cysteine, and threonine, which are prone due to their functional groups. The hydroxyl radicals formed by radiolysis react with the main protein chain, leading to oxidation and loss in activity of the enzyme (Eq. 8.4).

$$OH^\bullet + P\text{-}CONHCH(R)\text{-}P + O_2 \rightarrow P\text{-}CONH_2 + RCO\text{-}P + HO_2^\bullet \quad (8.4)$$

Apart from oxidation, the formation of dimers and trimers has been observed in some amino acids, such as tyrosine and its peptides. Irradiation also influences sulfhydryl enzymes, such as papain, lactate dehydrogenases, malate dehydrogenase, etc. Gamma irradiation delays fruit ripening by decreasing malate dehydrogenase activity (this enzyme is responsible for many metabolic reactions, including the citric acid cycle). This can be attributed to the destruction of cysteine and tryptophan residues in the active site and the enzyme's denaturation. Therefore, various organic species generated by radiolysis play various roles by attacking the amino acids, especially at active sites of the enzymes.

8.3 Effect of Irradiation on Nutritional Quality

Macronutrients are majorly unaffected by irradiation. Even with doses as high as 70 kGy, proteins' digestibility and biological value remain unaffected. When it comes to micronutrients, minerals majorly remain unaffected. Vitamins are pretty sensitive to irradiation practices. The loss of vitamins majorly depends on intrinsic factors such as food composition and extrinsic factors such as oxygen level, packaging, and temperature. Thiamine (Vit-B_1) is one of the most sensitive nutrients to irradiation. Meat, an essential source of B vitamins, may be significantly influenced by irradiation. However, cobalamin (Vit-B_{12}), niacin (Vit-B_3), and pyridoxine (Vit-B_6) remain almost unaffected. Packaging foods in a modified atmosphere under reduced oxygen can help minimize vitamin losses. Vitamins D and K are insensitive to radiation treatment. Vitamin E is more sensitive to irradiation than vitamin A; however, sources containing vitamin E such as vegetable fats are not commercially irradiated. In some cases, vitamin C undergoes degradation at high radiation doses. Irradiation also can induce oxidation, polymerization, dehydration and decarboxylation in fatty acids. Thus, food materials rich in lipids may develop off-flavor due to oxidation and lose certain fat-soluble nutrients. β-carotene, the precursor of vitamin A, is usually not affected significantly. However, the source of β-carotene and dosage play a significant role.

Irradiation can help degrade nitrites and nitrates significantly, added as additives to meat. The nitrites and nitrates are precursors to nitrosamines, which are potential carcinogens. Irradiation can also help to degrade the volatile nitrosamines. As already discussed, there can be radiolysis of triglycerides, resulting in 2-alkylcyclobutanones, which exhibit no mutagenicity and genotoxicity when consumed in low amounts. However, consumption of higher doses results in damage to human colonic cells.

8.4 Critical Processing Factors

8.4.1 Type of Product

The type of food matrix can be a critical deciding factor of the intensity of irradiation and the potential effects of irradiation on the food. For instance, higher moisture content can increase the probability of hydroxyl radical formation, leading to a more significant oxidative effect. Irradiation can also induce depolymerization of starch and pectin, leading to softening. High lipid content can also result in rancidity due to oxidation induced by radicals generated by ionizing irradiation. Usually, low doses of 0.05–0.2 kGy are enough to inhibit sprouting in potato tubers, but higher doses up to 1 kGy can be used. Higher doses of 1 kGy are not usually recommended for sprouting inhibition. They may also lead to enhanced susceptibility of pathogens and an increase in reducing sugar contents of the tubers. It is also an alternative to fumigation, as it destroys all the insects, and thus can used for pest management (may require 0.1–0.5 kGy). This is facilitated usually at doses up to 0.75 kGy. Irradiation in the range of 0.25–1 kGy can also delay the ripening of fruits and vegetables, along with the reduction of radiosensitive spoilage microbes in meat, fish, and poultry. Intermediate doses (1–10 kGy) can work efficiently with spices, herbs, and dry cereal powders. Low acid foods, which need to achieve commercial sterility and be on the shelf under ambient conditions, may require high doses of 25 kGy or even greater. Hospital diets, usually given to immunecompromised patients, should be sterile and can be given doses up to 70 kGy.

The chemical reactions and oxidative damages can also lead to loss of antioxidants, off-flavor generation, lipid oxidation, and change in color of food items, leading to a decrease in sensory quality. Due to the radiolysis of water playing a major role in the lethality of microbes and the possibility of lipid oxidation, the moisture content, lipid content, and antioxidants of foods play a major role. The physical state, i.e., the food in frozen form, is more resistant than unfrozen foods. This may be due to the availability of lower aqueous water in frozen foods. Other food components than water also influence the decimal reduction (D_{10}) values of microorganisms and enzyme activity. The complexity of the food matrix may protect the microbe, thus increasing the D_{10} (irradiation dose required to achieve 1 log reduction) values.

Another intrinsic factor that can contribute to the radiolysis of water is the pH of the food, which can influence the protonation and deprotonation of various molecular species. Thus, pH, oxygen level, and other food components may also influence the inactivation of microorganisms and enzymes. As an extrinsic factor, temperature influences the *indirect* effects more than the *direct* effects. As the temperature decreases, a lower molecular level movement reduces the probability of the formation of *indirect* radiolytic products.

8.4.2 *Target Microorganism or Enzyme*

The most resistant microflora or enzyme in a particular food product is chosen as the target entity to ensure adequate inactivation of other microbes and spoilage enzymes. Different biological entities show different susceptibility to the variable dose levels (**Fig. 8.4**). The lethal dose for humans is around 10 Gy. Food spoilage bacteria in fresh produce such as mangoes, berries, and other fruits can be destroyed by intermediate doses (1–10 kGy). It can also destroy foodborne pathogens, such as *Salmonella* in frozen meats, poultry eggs, and shrimps. Non-spore-forming pathogenic bacteria of public health concerns, such as *Listeria*, *Yersinia*, and *Campylobacter*, can be effectively controlled at lower doses. Microbes, which survive this treatment, become more susceptible to further processing steps if any. Therefore, a combination

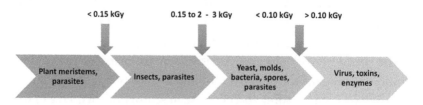

Figure 8.4 Susceptibility of different biological targets against different dose levels of irradiation.

of intermediate doses and other treatments can be used to achieve the required lethality. Gram-positive bacteria have thicker peptidoglycan layer in the membrane which possibly imparts higher resistance than Gram-negative bacteria. As already discussed, viruses are highly resistant to irradiation, and high-dose treatments can ensure the reduction of viral contamination. There is limited literature on the susceptibility of spoilage enzymes toward ionizing irradiation, and therefore it becomes difficult to comment on the order of sensitivity of various enzymes. However, enzymes and viruses are the most resistant entities, requiring more than 10 kGy of ionizing irradiation for adequate inactivation.

The radiosensitivity of various microbial species and enzymes plays a vital role in determining the process parameters of irradiation of foods. However, some bacteria, such as *Deinococcus* species, are highly resistant to ionizing radiation due to a highly efficient DNA repair system. Efficient nucleic acid repair mechanisms, a complicated cell envelope, and carotenoids in *Deinococcus radiophilus* have been postulated for their resistance towards radiation. The Gram-negative bacteria and aerobic bacteria are more sensitive to radiolysis than Gram-positive bacteria and anaerobic bacteria, respectively. Products having lower water activity have lesser *indirect* effects induced by radiation. This also explains the higher stability of spore-formers toward irradiation treatment because they can grow in low water activity containing food.

8.4.3 Irradiation Equipment and Process Parameters

The various irradiation sources, which are approved for food processing, are as follow:

  a. γ-irradiation of 0.66 MeV using Cs-137 radioisotope.
  b. γ-irradiation of 1.17–1.33 MeV using Co-60 radioisotope.
  c. Electron beams (cathode rays) from machine sources, with energies in the range of 0.1–10 MeV.
  d. Electron (beta rays) from radioisotope possess energy 0.01–1 MeV.
  e. X-rays from machine sources carry energy in the 0.01–10 MeV range.

The various radioactive sources, such as Caesium (atomic number 55 and mass 137) and Cobalt (atomic number 27 and mass 60), do not have any issues with radioactive waste disposal, as they decay over the years to nonradioactive Caesium (30.17 years half-life for Cs-137) and Cobalt (5.26 years half-life for Co-60). In addition to the major gamma emission, X-rays and electrons are also emitted by Co-60 and Cs-137. X-rays and electron beams can be generated using an *electron accelerator* (**Fig. 8.5b**). In other words, the electron beam can be converted into an X-ray by causing the energetic electrons to strike any metal target (acting as anode). The essential components of an *electron accelerator* include an electron source (cathode), an accelerator tube, and a beam shaping system. The cathode (e.g., tungsten made) that is electrically excited under vacuum conditions emits electrons that are carried up and exposed to a potential difference in an electric gun. The electrons are accelerated to energy proportional to the applied voltage and then shaped into a beam, which will be pointed toward the

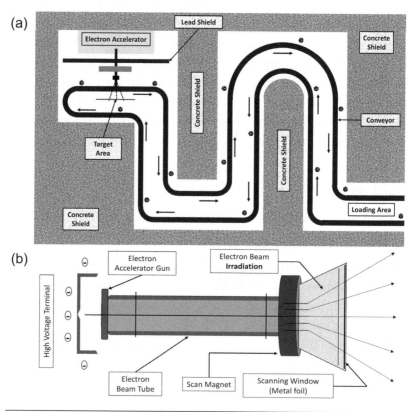

Figure 8.5 Electron accelerator (can produce both electron beam and X-ray) based (a) plant layout for irradiation of food products and (b) diagram of a constant potential, scanned beam electron accelerator. (Adapted from Ortega-Rivas (2012) and Marshall (2012))

food sample to be irradiated. **Fig. 8.5a** shows a layout of an electron beam or X-ray treatment plant. The gamma irradiators are encapsulated in stainless steel tubes, making them doubly safe from any kind of leakage of irradiation. Gamma-irradiation and X-rays have higher penetration power and can be used in relatively thick and dense fruit products. However, electron beams can be used when the shallow conversion can be enough for the required use. None of the limits prescribed for irradiation induce any kind of radioactivity in food, making it further safe to consume.

The major components of an irradiation equipment system include the irradiation source, the irradiating cell, the concrete wall protection system, and the automatic conveyor. The automatic conveyor passes through the irradiation chamber with the prepacked food material. The radiation chamber is covered with a 1.5–1.8 m layer thick concrete wall, which prevents any kind of leakage of irradiation into the working area and poses no risk to the operators.

The radiation source is stored underwater in an underground facility when not in use, with water and underground set-up acting as a shielding source. This is majorly applicable to γ-irradiation, where radionuclide sources are required (**Fig. 8.6**). When the irradiation starts, the radionuclide source is lifted from its position. This flexibility of wet storage makes it feasible for the operator to undertake plant maintenance. However, electron beam irradiation, which uses a machine source instead, can be switched 'ON' and 'OFF', according

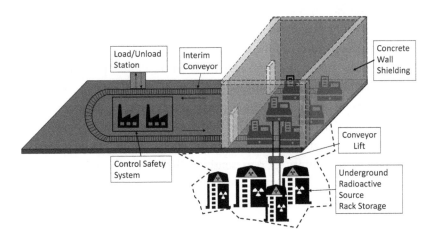

Figure 8.6   Gamma irradiation facility for food materials.

to the need. Apart from the above parts of the equipment, safety inter-locks alarms should also be in place, which will maintain control to access and indicate the radiation level, water level, and the position of the food product in the irradiation cell. The radiation facilities can be usually divided into research, pilot, and commercial facilities. Some of the critical parameters are the radioactivity of the source, the physi-cal properties of the food, and the food application, which may range from sprout inhibition to sterilization of food.

The primary process parameters include:

i. *Absorbed dose* (D) refers to the amount of radiation energy absorbed per unit mass of food, expressed in Gy. It is described as the mean energy E of irradiation imparted to the unit mass of a food product. To measure the overall average dose, dose meters or dosimeters are positioned at different points of the treatment chamber. There are various dosimeters but choosing an appropriate dosimeter system for the particular irradiation set-up is crucial.

ii. *Absorbed dose rate* ($\dot{D}$) is the rate of change of absorbed dose with time.

iii. *Throughput* is the amount of food material or the output pro-cessed per unit time. Gamma irradiation is used for small-throughput processes, and electron beams are majorly used for high throughput processes, where shallow penetration is enough for the product.

iv. *The thickness of the food product* is another crucial component of irradiation. The depth-dose distribution helps to understand the variation of dose and the thickness of the product. In such cases, the dose uniformity ratio (*DUR*) is an important parameter to understand the maximum and minimum doses experienced by the sample (Eq. 8.5). *DUR* is usually ≥ 1, and it typically increases with the density, volume, and thickness of the product, showing more significant non-uniformity in treatment. *DUR* is expected to be close to unity for representing treatment uniformity.

$$DUR = \frac{D_{max}}{D_{min}} \tag{8.5}$$

v. *Change over time* is the time required to change over to a dif-ferent product with varying densities.

8.5 Dose of Irradiation

Decimal reduction dose (D_{10}) is the dose of ionizing radiation responsible for inactivation or reduction of 90% (1 log cycle) of the initial microbial load or enzymatic activity. While D_{10} is measured only for microbes and enzymes, this concept can be extended to nutrients and bioactive compounds. Conventional first-order decay usually explains the amount of irradiation required to bring about a required microbial inactivation. One of the central assumptions of this model is that a single target inactivation is sufficient to bring about the inactivation of the microbes. This is also known as the single hit theory model (Eq. 8.6).

$$\ln\left(\frac{N}{N_o}\right) = -k \cdot D \qquad (8.6)$$

N_o is the initial microbial load, and N is the microbial load after an absorbed dose D (Gy) with an inactivation rate of k (Gy^{-1}, for Eq. 8.6) concerning the applied dose. While a single hit theory may be applicable and work well with modeling some microbes, the multi-target theory assumes n number of identical targets, which are targets for irradiation. Probability theory can be applied to this model, resulting in Eq. 8.7.

$$N/N_o = 1 - (1 - e^{-k.D})^n \qquad (8.7)$$

The dose delivered to achieve a particular effect, either sterilization or inhibition of sprouting, is optimized after various studies on a specific type of food matrix. The *DUR* increases in products with greater thicknesses and larger containers. This can be brought close to one by irradiating products from both sides, rather than one. Other methods could include the rotation of the product container or changing the product's position with respect to the source. Thus, the choice of the dose would depend on the objective of the treatment, i.e., decontamination, pasteurization or sterilization, the decimal reduction dose of the target microorganism, the initial load, and the efficacy of the treatment.

8.6 Practical Example

One study on the decontamination of prepackaged powdered spices such as pepper, turmeric, chili, and coriander using γ-irradiation will make it clearer (Munasiri et al., 1987). The heaviest load of natural bacteria was found in chili, followed by pepper, turmeric, and coriander. The natural fungal load was also maximal in chili, followed by coriander, turmeric, and pepper. The initial bacterial and fungal counts in spices were 10^5–10^7 and 10^2–10^6 CFU/g. *Irradiation treatment:* The spices have received an average absorbed dose of 5, 10, and 12.5 kGy with a maximum dose rate of 27 Gy/min. The total bacterial and fungal counts were detected, along with an enumeration of *Bacillus cereus, Bacillus stearothermophilus, Staphylococcus aureus, Streptococci, Lactobacilli,* and *Salmonella.* The bacterial population was composed majorly of spores.

Observations: A dose of 10 kGy was sufficient to reduce all bacterial counts to zero; a dose of 5 kGy was sufficient for complete mold inactivation. There were no significant changes in the color (as per sensory panel scores) or the sensory perception of pungency in the spices during storage after treatment at ambient temperature. This was well correlated to the minimal changes in curcumin and capsanthin content of spices. Only a 4.7% reduction in color was observed in irradiated spices at 10 kGy after 8 months of storage in ambient conditions. It can be concluded that molds were more easily inactivated than spore-forming bacteria by γ-irradiation.

8.7 Challenges

Radiation processing is not suitable for processing all food products, particularly in high-lipid foods. Therefore, dairy-based products are not suitable for irradiation. Irradiation system typically requires extensive facilities with high investment. Some pathogens are highly resistant to irradiation. Microbes such as *Clostridium botulinum* and viruses, and yeast molds are highly resistant to irradiation. It is often required to deliver high doses to breakdown the toxins. Any food product processed by irradiation requires clear labeling and indication on the label. Adequate throughput must be maintained to maximize the utilization of the cost of operations. The setting of an irradiation plant requires approvals from regulatory organizations in the specified country. The operators also have to take extra precautions for safety, which may add to the cost.

Irradiation is also losing momentum due to the stereotype of the word 'radiation' in consumers' minds. The labeling of the foods as irradiated seems more like a warning symbol than information about the product to the consumer. Consumer education and awareness regarding its advantages and safety assurance are required for the growth of the market of irradiated foods. Some fruits are highly radiosensitive and may undergo a loss of firmness and development of metabolic disorders. Climacteric fruits usually experience these issues, where irradiation acts as abiotic stress. Sometimes, irradiation may also induce depolymerization of pectins, which can decrease their gelling power too. Hence, dose-response studies must be done accurately for a specific food product before scaling up. This should include examining the physical, microbial, and sensory properties of foods.

8.8 Summary

- Ionizing irradiations including electron beam, X-ray, and gamma rays are effective non-thermal techniques for the treatment of fresh produce, poultry, meat, and seafood.
- Photons can interact with any matter in three ways, *Compton effect*, *pair production*, and the *photoelectric effect*.
- Irradiation can attain microbial inactivation by attacking cell's genetic materials and organelles both *directly* with intense ionizing radiation and *indirectly* with radicals, hydrogen peroxide, and ozone formed via radiolysis of available water.
- Ionizing irradiation does not cause any significant effect on the nutritional content of food, with little effect on sensory properties and treated products are safe to consume.
- Viruses and enzymes are the most resistant entities to ionizing irradiation and may require > 10 kGy doses.
- A significant challenge is the depth dose distribution and the density of food product, impacting the quality of irradiated food.
- Consumers need to be made aware and convinced regarding the safety of irradiated foods for the growth of the market of irradiated foods.
- Irradiation can be a challenge to various categories of foods. Therefore, dose-effect relationship and shelf-life studies are necessary to make a decision.

- Training of personnel and safety precautions are required for irradiation processing of foods.
- Irradiation may have detrimental effects on food quality, such as oxidation of lipids, depolymerization of pectin, and oxidation of proteins.

8.9 Solved Numerical

1) A Co-60 irradiator is pasteurizing frozen seafood at a minimum dose of 6 kGy, with the power of the irradiator being 15 kW. Considering the maximum efficiency, calculate the throughput of the irradiator. After the first shift, the second shift needs to sterilize beef carcasses at a minimum dose of 40 kGy. (a) Calculate the change in throughput due to this changeover of the sample. (b) The Co-60 irradiator is also known to have a 12% reduction in activity annually. Calculate the change in throughput that needs to be made after a year of operation to maintain the same dose absorbed by seafood and frozen beef.

Solution:

$$\dot{m} = \varepsilon \times \frac{P \times 3600}{D}$$

Where ε is the radiator's efficiency, P is the power of the irradiator in kW, D is the absorbed dose in kGy, and $\dot{m}$ is the throughput in $kg \cdot h^{-1}$.
   For frozen seafood; $P = 15$ kW, $D_1 = 6$ kGy, $\varepsilon = 1.0$
   From the above equations and substitutions,

$$\dot{m}_1 = 1.0 \times \frac{15 \times 10^3 \times 3600}{6 \times 10^3}$$

$$\Rightarrow \dot{m}_1 = 9 \times 10^3 \; kg \cdot h^{-1} = 9 \; t \cdot h^{-1} \; (\text{tonnes per hour})$$

For beef carcass, $P = 15$ kW, $D_2 = 40$ kGy, $\varepsilon = 1.0$

$$\dot{m}_2 = 1.0 \times \frac{15 \times 10^3 \times 3600}{40 \times 10^3} \qquad \Rightarrow \dot{m}_2 = 1.35 \; t \cdot h^{-1}$$

a. When sample changeover is done from seafood to beef carcasses, the throughput decreases from 9 to 1.35 $t \cdot h^{-1}$, representing a decrease of 7.65 $t \cdot h^{-1}$.

b. Assuming that a 12% decrease in the activity of the Co-60 source will directly correlate to a 12% reduction in power of the irradiator, the throughput needs to be reduced so that the same amount of dose is received by the food materials. After a year, the throughput of frozen seafood and beef carcass would be reduced by 12% as well, which can be calculated as 7.9 t·h⁻¹ and 1.19 t·h⁻¹, respectively.

2) The following equation gives the depth dose distribution of beef fillets of 4 cm thickness:

$$D = x^2 - 4x + 18$$

Where x is the thickness of the fillet and D is the irradiation dose. It is also shown in **Fig. 8.7**. Calculate the log reduction in microflora at the lowest dose point, considering the rate of inactivation (k) to be 0.25 kGy⁻¹ as per the first-order kinetics. The irradiation system was supplied with a power of 15 kW source and functioned at 80% efficiency. Estimate how many fillets can be processed if the average weight of a single fillet is 150 g? The maximum *DUR* for the beef was tolerable up to 1.5. Comment whether the dose would be suitable for the required purpose.

Solution: The dose (D) at $x = 0$ and $x = 4$ corresponds to 18 kGy, the maximum dose received by the fillet. For minimum dose calculations, the local minima need to be calculated.

$$\frac{dD}{dx} = 0; \quad \frac{d^2 D}{dx^2} > 0$$

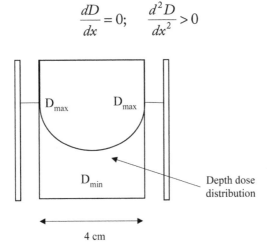

4 cm

Figure 8.7   The schematic showing the summary of the numerical problem Q2.

Single differentiations of the quadratic equation of dose gives: $\frac{dD}{dx} = 2x - 4$ and double differentiation of the quadratic equation of dose gives; $\frac{d^2D}{dx^2} = 2 > 0$

From the above equation, $x = 2$ comes out to be the point of local minima, where the minimum dose is 14 kGy as per the given quadratic equation.

Assuming first-order kinetics,

$$\log\left(\frac{N}{N_0}\right) = \frac{-k \cdot D}{2.303}$$

Where, $D = 14$ kGy, $k = 0.25$ kGy^{-1} and

$$\log\left(\frac{N}{N_0}\right) = \frac{-0.25 \times 14}{2.303} \Rightarrow \log\left(\frac{N}{N_0}\right) = -1.52;$$

therefore, the log reduction observed is 1.52 at the lowest dose.

Considering $P = 15$ kW, $D = 18$ kGy and $\varepsilon = 0.8$, the throughput;

$$\dot{m} = 0.8 \times \frac{15 \times 10^3 \times 3600}{18 \times 10^3} \Rightarrow \dot{m} = 2.4 \ t \cdot h^{-1}$$

Having a fillet of 150 g, the number of units that can be processed is 16,000. The *DUR* is defined as the ratio of the maximum dose to the minimum dose experienced by the product;

$$DUR = \frac{Max.D}{Min.D} = \frac{18}{14} = 1.29$$

The value of *DUR* should always be close to unity to maintain treatment uniformity and to achieve a particular lethality in the product. So, treatment suitability can be confirmed if $1 \le DUR \le 1.5$. Thus, $DUR = 1.29$ indicates that current irradiation is suitable for the purpose.

3) An unshielded X-ray source is shielded using lead, having an absorption coefficient (μ_{Pb}) of 0.77 cm^{-1}. The shield experiences an emission of 750 W·m^{-2} at 1 m distance from the source (**Fig. 8.8**). Calculate the half-value layer (HVL) of the shielding material. Also, calculate the change in HVL if the lead is replaced with iron ($\mu_{Fe} = 0.44$ cm^{-1}).

Solution: The expression gives the absorbed irradiation in the above figure, showing $I = I_0 \cdot e^{-\mu \cdot d}$, where '$I_0$' is the initial incident intensity of irradiation that attenuates or diminishes (following an exponential

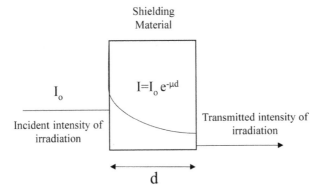

Figure 8.8 The schematic showing the summary of the numerical problem Q3.

decay) up to an intensity 'I' as it propagates along with the thickness or path length (d) of the medium/material.

HVL is the thickness of the medium/material causing the incident intensity to reduce by half; $I = I_0/2 = 750/2 \ \text{W·m}^{-2} = 375 \times 10^4 \ \text{W·cm}^{-2}$

$$I_0/2 = I_0 \cdot e^{-\mu_{Pb} \cdot d_{Pb}} \qquad \Rightarrow 0.5 = e^{-0.77 \times d_{Pb}}$$
$$\Rightarrow \ln(0.5) = -0.77 \times d_{Pb} \qquad \Rightarrow d_{Pb} = 0.90 \ \text{cm}$$

So, HVL (d_{Pb}) of lead will be 0.90 cm. Similarly, in case of iron, the HVL will be:

$$\ln(0.5) = -0.44 \times d_{Fe} \Rightarrow \text{HVL} \ (d_{Fe}) = 1.56 \ \text{cm}$$

Lead (Pb) has a much higher radiation absorption ability than iron (Fe).

8.10 Multiple Choice Questions

1. Which of the following irradiations can be used for high throughput operations?
   a. Gamma Irradiation
   b. Electron Beam Irradiation
   c. X-Ray Irradiation
   d. Both a & c

2. Which of the following are destroyed by gamma irradiation?
   a. Pesticides
   b. *Staphylococcal enterotoxin*
   c. Aflatoxin
   d. Viruses

3. What is the typical dose range of pest management?
    a. 0.05–0.18 kGy
    b. 0.1–0.5 kGy
    c. 1–5 kGy
    d. None of the above

4. Which extrinsic factors positively affect the indirect pathway for microbial or enzyme inactivation through irradiation?
    a. Oxygen
    b. Water
    c. Temperature
    d. All of the above

5. Which of the following can be the most sensitive to ionizing irradiation? (Hint: Differentiate in terms of Gram-negative, Gram-positive, and fungi)
    a. *Pseudomonas*
    b. *Staphylococci*
    c. *Aspergillus*
    d. *Micrococci*

6. The amino acid most susceptible to electron beam irradiation is
    a. Methionine
    b. Cysteine
    c. Leucine
    d. Arginine

7. Which of the following effects are not possible by ionizing irradiation?
    a. Degradation of pectin and cellulose
    b. Modification of viscosity of albumin solutions
    c. Development of rancidity in foods
    d. None of the above

8. _____ is the most dominant mechanism of photon interaction and any tissue type material. This mechanism involves deflecting a photon through a certain angle as the photon collides with an orbital electron of the atom of the material, transferring some of its energy and causing the electron to get ejected out of the orbit.
    a. Pair production
    b. Photoelectric effect

c. Compton scattering
d. Deflective photon collision

9. To confirm uniform treatment of irradiation and process adequate lethality, the *DUR* is an essential parameter; its value usually exists as:
   a. $DUR \geq 1$
   b. $DUR \leq 1$
   c. $-1 \leq DUR \leq 1$
   d. $0 < DUR \leq 1$

10. Which isotope of Cobalt (Co-'atomic mass') is usually used for gamma-ray productions due to its relatively high stability?
    a. Co-47
    b. Co-54
    c. Co-60
    d. Co-75

8.11 Short Answer Type Questions

a. Discuss the mechanism of the inactivation of enzymes by ionizing irradiation.
b. How is ionizing radiation different from nonionizing radiation, and why is ionizing irradiation also known as cold sterilization?
c. List the various types of ionizing irradiation used in food irradiation, with their sources.
d. What are the mechanisms of energy loss by a photon when it interacts with any material? Elaborate with a schematic diagram.
e. Explain the difference between the direct and indirect effects of irradiation.

8.12 Descriptive Questions

a. Explain the effect of ionizing irradiation on the nutritional quality of food.
b. Discuss the challenges to the use of ionizing irradiation, with its limitations.

c. Explain the process of irradiation with the equipment required and various process parameters.

d. Comment on the susceptibility of various classes of microorganisms to ionizing irradiation.

e. Elaborate on the intrinsic and extrinsic factors, which can affect the efficacy of ionizing irradiation treatment.

8.13 Numerical Problems

1) Gamma irradiation has been used to inactivate lipase enzymes in wheat germ. It has been observed that lipase activity was 82% after treating with 12 kGy irradiation dose and 69% after receiving a dose of 30 kGy. Assume the initial activity of lipase as 100%. (a) Develop a first-order kinetic model (based on irradiation dose), (b) calculate the sum of the square of errors to confirm the accuracy of the model, and (c) estimate the predicted enzyme activity from the model at a dose of 20 kGy.

2) Effect of gamma irradiation was tested for inactivating *Staphylococcus aureus* inoculated on the surface of the grounded chicken. The inoculation size of *S. aureus* was 10^7 CFU·g^{-1}. The log count ($\log(N)$) of *S. aureus* was analyzed at different irradiation doses (D, kGy), and a quadratic polynomial model was developed; $\log(N) = 0.14D^2 - 2.03D + 7.06$. Estimate the log reduction after treating with 2 kGy and 4 kGy. Assume the irradiation system efficiency to be 80%. What will be the throughput (kg·h^{-1}) if the system consumes 15 kW power and the chicken is exposed to an irradiation dose of 4 kGy. Find out how many samples of grounded chicken can be processed in an hour, with each sample weighing about 50 g?

References

Munasiri, M. A., Parte, M. N., Ghanekar, A. S., Sharma, A., Padwal-Desai, S. R., & Nadkarni, G. B. (1987). Sterilization of ground pre-packed Indian spices by gamma irradiation. *Journal of food science, 52*(3), 823–824.

Suggested Readings

Ferreira, I. C. F. R., Antonio, A. L., & Verde, S. C. (2018). *Food Irradiation Technologies: Concepts, Applications, and Outcomes.* Royal Society of Chemistry, Piccadilly London, UK.

Marshall, R. C. (2012). Advances in Electron Beam and X-ray Technologies for Food Irradiation. In X. Fan and C. H. Sommers (Eds.), *Food Irradiation Research and Technology*. Wiley-Blackwell, Chichester, West Sussex, UK. pp. 9–28.

Munir, M. T., & Federighi, M. (2020). Control of foodborne biological hazards by ionizing radiations. *Foods (Basel, Switzerland)*, 9(7), 878.

Ortega-Rivas, E. (2012). Ionizing Radiation: Irradiation. *In Non-thermal Food Engineering Operations*. Springer, New York, USA. pp. 231–248.

Ravindran, R., & Jaiswal, A. (2019). Wholesomeness and safety aspects of irradiated foods. *Food Chemistry*, *285*, 363–368.

Saha, A., Mandal, P., & Bhattacharyya, S. (1995). Radiation-induced inactivation of enzymes—A review. *Radiation Physics and Chemistry*, *46*, 123–145.

Woodside, J. (2015). Nutritional aspects of irradiated food. *Stewart Postharvest Review*, *11*, 1–6.

Answers for MCQs (sec. 8.10)

1	2	3	4	5	6	7	8	9	10
d	c	b	d	a	b	d	c	a	c

9

OSCILLATING MAGNETIC FIELD PROCESSING

9.1 Principle and Operation

Magnetism is the force of attraction and repulsion exerted among magnets and magnetically active components. An alternating or direct current (AC/DC) in conductors or permanent magnets generates a magnetic field (MF) (**Fig. 9.1**).

Electromagnetic waves are produced by an oscillating charge, which is in the form of oscillating electric and magnetic fields mutually perpendicular to each other (**Fig. 9.2**). Their motion is represented by the Poynting vector (quantity expressing the magnitude and flow of energy in an electromagnetic wave). Based on their behavior over time, they are categorized as static magnetic fields (SMFs) and oscillating magnetic fields (OMFs), also known as pulsed or alternating magnetic fields (PMF or AMF). The spatial distribution of an MF will be homogeneous if the field gradient is zero (in uniformly distributed MF, a change in the magnitude of MF for a change in distance will remain zero) over the space where samples are exposed; otherwise, it will be heterogeneous. Based only on magnetic flux density (denoted as $\vec{B}$, expressed in 'Tesla', T), SMFs are categorized as follows:

 a. Super weak (100 nT to 0.5 mT)
 b. Weak (< 1 mT)
 c. Moderate (1 mT to 1 T)
 d. Strong (1–5 T)
 e. Ultra -Strong (> 5 T)

The magnetic flux (φ, expressed in weber 'Wb' or voltage·second 'V·s'), magnetic flux density (B, expressed in weber per unit area 'Wb·m^{-2}' or tesla 'T'), MF strength or magnetic force (H, expressed in ampere per unit length 'A·m^{-1}'), magnetization (M, shares the same unit as H), frequency (ν, expressed in Hz), waveform, exposure

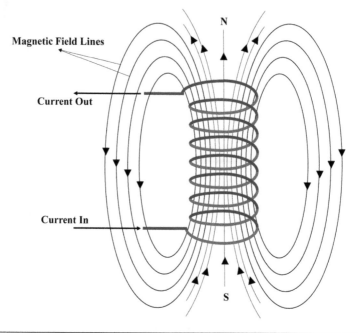

Figure 9.1 Magnetic field generated when current is passed through a metallic coil.

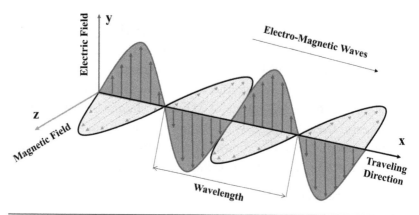

Figure 9.2 Propagation of an electromagnetic wave with a magnetic field perpendicular to the electric field.

duration (time), and polarity are the characteristics evaluated for the categorization of OMFs and PMFs. They are considered in terms of frequency as follows:

a. Extremely low frequency (0–300 Hz)
b. Intermediate frequency (300 Hz to 1 MHz)
c. Radiofrequency (1–500 MHz)

d. Microwave frequency (500 MHz to 10 GHz)

e. High frequency (> 10 GHz)

Higher energy OMFs may be utilized in non-thermal food preservation methods. It is used to inactivate bacteria and denature spoilage enzymes in fresh or minimally processed foods. OMF has many applications in food processing, including aseptic treatment of solid and liquid foods within a compact package, minimal heat generation inside the food, and lower energy requirements. OMF processing of roast beef, milk, and juices are also possible food applications.

9.2 Mechanism of Lethality

MFs are increasingly being utilized in food processing to maintain food quality. Depending on the equipment, they can be static (SMF) or oscillating (OMF). Various processes have been proposed to explain the impact of MFs on foods such as the radical pair and cyclotron resonance processes imply quantum correlations at the subatomic particle level.

9.2.1 Microbial Inactivation

The MF's impact on microbial growth is sometimes conflicting. MFs and the produced electric fields can either hinder or encourage the development of microorganisms under exposure. Induced electric field strength and induced current density quantify the electric field induced in the microbial system as a result of absorbed MFs. The major objective of using MFs on microorganisms is to minimize pathogen load. However, OMF can also increase the growth of certain good bacterial strains to ferment food. According to the researchers, an MF works on bacterial cell degradation, preventing them from growing and reproducing. MFs with low frequencies have a major influence on cells and tissues. Low frequency and/or high-intensity MFs damage cell membranes, organelles, and DNA of bacteria. It has been observed that magnetic flux concentrations of 5–10 Tesla (T) and frequencies of 5–500 kHz are required to inactivate microorganisms.

There exist many theories regarding MF-based microbial inactivation. The popular ones are models related to the movement of ions (naturally present in a cell or its surrounding) under the influence of MF.

Two alternate theories are the radical formation and transfer of coupling energy from MF to the magnetically active components of cell organelles. These are responsible for the damages to various cell organelle, membrane, and essential metabolic intermediates.

The first theory about the translocation (moving from one place to another) of ions under MF explains that low-intensity OMF (similar to the MF of Earth) can weaken the ion-protein bonds. Many vital cellular proteins of microorganisms contain ions. The biological effects due to steady MFs are more prominent around specific frequencies, the cyclotron resonance frequency of ions. Thus, the theory is called the Ion Cyclotron Resonance (ICR) model. Cyclotron is a type of particle accelerator; its name is used to represent the movement of ions under MF. Subsequently, the term resonance means that when the frequency of periodically applied force is the same or close to the natural frequency (resonant frequency) of a dynamic system (wave), the amplitude of the oscillating wave is enhanced. When an ion having charge q enters an MF (B) at velocity v, it experiences a force F (Eq. 9.1).

$$F = q\vec{v} \times \vec{B} \tag{9.1}$$

Eq. 9.1 shows that F will be zero if v and B are parallel (because there exists a cross product between v and B), but if the ion's velocity is normal to the MF (B), the ion takes a circular path due to the Lorentz force. The ions revolve in an applied MF at a frequency called gyrofrequency or cyclotron angular frequency ($\omega = 2\pi f$, where f is the linear form of frequency) that depends on the ratio of charge (q) and mass (m) (Eq. 9.2).

$$\omega = \frac{qB}{m} \quad => f = \frac{qB}{2\pi m} \tag{9.2}$$

Cyclotron resonance occurs when f becomes equal to the frequency of the applied MF. **Fig. 9.3** shows the movement of a charged particle in an MF. For example, the resonance frequency of Na^+ is 33.33 Hz, and Ca^+ is 38.7 Hz under an MF intensity of 50 μT. At cyclotron resonance, energy is transferred selectively from MF to the ionic species. Ions are the interaction site of MF (ions are forced into motion by MF) within the cells of the microorganism, where the effects of MFs are transmitted to the other cell organelles, other cells, and

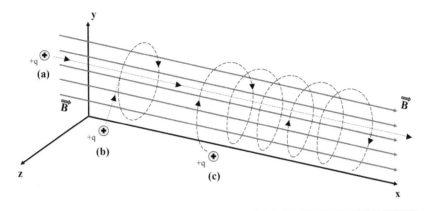

Figure 9.3   Movement of a charged particle (q) at a certain velocity (v_q) in the flux of magnetic field (B), when a particle enters with (a) v_q parallel to B, (b) v_q perpendicular to B, and (c) v_q arbitrary angle to B.

tissues. This also indirectly disrupts the metabolic activities vital for survival.

The second theory about the translocation of ions under MF is the Ion Parametric Resonance (IPR) model. The parametric resonance occurs when a mechanical system is parametrically excited with an external frequency equivalent to twice the natural frequency. Parametric excitation is different from regular resonance as the action appears as a time-varying modification on a system parameter. When ions like calcium ion experiences a weak external DC magnetic field (SMF both in terms of magnitude and direction) modulated with an additional AC magnetic field (AMF) or OMF in parallel to the SMF, the ion follows a motion described as a coherent combination of circular orbits. More importantly, IPR considers how ion cofactors (nonprotein components or metallic ions required for an enzyme to act as a catalyst) that are part of the key molecular complex such as enzyme binding site, initiate conformational changes of a molecular complex that can further cause observable changes to the biological system. Thus, the metabolic reactions and their reaction rate influenced by ions can be controlled by restructuring the internal energy state with the help of the externally applied MF. The effectiveness of the restructuring process affecting ions depends on factors, such as the frequency of applied AC-MF, the flux density of applied DC-MF, the charge-mass ratio of the specific ion, and its respective spatial properties to allow sufficient resonance time to initiate any effect. For instance, calmodulin is a calcium ion-bound protein. The calcium ion continually vibrates across equilibrium positions at the binding site of calmodulin protein. The exposure of SMF causes the plane of vibration to rotate in the direction of the MF at exactly half of the cyclotron frequency of the bound calcium. Overall the application of OMF at calcium cyclotron frequency disturbs the calcium-protein bond.

Sometimes the growth of microorganisms can be guided by low-intensity MFs, such as the Earth's MF (0.305×10^{-4} T). For instance, budding in *Saccharomyces cerevisiae* yeast was suppressed in cultures subjected to MFs. Alternate theory about the movement of paramagnetic (components that are weakly affected by MF under vacuum than in standard conditions) free radicals to the area with the highest field strength shows that these free radicals are responsible for producing metabolic disturbances and damage to DNA.

The second theory dictates that biomolecules are anisotropic by nature, which allows the biological membranes to exhibit strong orientations, such as parallel or perpendicular orientation to the applied MFs, resulting in peptide bond resonance. The theory suggests that OMF may couple energy into the magnetically active component of large vital molecules, such as DNA. In mechanical physics, a couple means a pair of equal and parallel forces acting in opposite directions, which creates rotation in a system without any acceleration of the center of the mass. MF in the range of 5–50 T can provide an energy of 10^{-2}–10^{-3} eV per oscillation to 1 dipole in DNA. Finally, numerous oscillations and collective assembly of sufficient local activation may lead to the breakdown of covalent bonds in DNA and inactivate the microorganism.

9.2.2 Spoilage Enzymes Inactivation

Enzymes are proteins that perform a specific function that can be helpful or harmful to the quality of food. Choosing the action levels (related theories or models of mechanism) of an MF would thus be decided by whether we intend to stimulate or inhibit the enzymes. MFs may change the structure of enzymes, altering biological processes. One research detected changes in four secondary conformations of α-amylase when SMF was applied. As discussed in the previous section of microbial inactivation, translocation of ions under MF (includes ICR and IPR theory) can initiate conformational changes and disrupt various ion-dependent reactions by damaging ion-enzyme complex and the bonds of essential ion cofactors with enzyme binding site. The formation of free radicals under the exposure of MF can also contribute to the inactivation of enzyme proteins. An MF can generally affect biological processes involving more than one unpaired electron. For example, the inactivation of peroxidase (POD) from horseradish after OMF treatment happened due to conformational changes. These were presumed to be the consequence of the unfolding of the native structure and exposing the aromatic amino acid residues previously buried in its natural compact globular structure to a relatively more polar environment.

On the other hand, the relative velocity of enzymes and substrate molecules must be synchronized to improve enzymatic activity. The mobility of enzymes may be controlled by immobilizing them with magnetic particles and regulating their movement with the frequency utilized in the MF. An MF with a frequency of 5 Hz can increase α-amylase activity.

9.3 Effect of OMF on Nutritional Quality and Freezing of Food Products

Water, proteins, carbs, fiber, lipids, and minerals are the primary components of food, which influence its physicochemical characteristics. However, except for the water that has evidence of MF effects, research on the application of MFs to foods has been characterized as a black box system until now. A black box system represents a process where input is converted into an output without knowing the governing mechanisms. Water is one of the primary constituents of any biological system or food product, which affects the interaction of the food material with MF. This interaction may enhance proteins' chemical activity or hydration of proteins and cellular structures.

When an MF is applied, the food component's behavior changes based on its isotropy and magnetic susceptibility χ (if $\chi < 0$, it is diamagnetic, otherwise paramagnetic). Isotropic susceptibility means the degree of magnetization is equal in all three orthogonal directions, and it does not show strong orientations under MF. Anisotropy is the opposite of isotropy. MF repels diamagnetic materials due to the formation of induced MF in the opposite direction of the applied MF. In contrast, paramagnetic materials are weakly attracted by MF. Carbon atoms are isotopically susceptible, whereas organic compounds in food are anisotropically susceptible. As a result, OMF uniquely affects different components of the food product. Thus, the properties of food molecules and compounds should be computed to better understand the impact of MFs on food physicochemical properties. For example, the 8.2% difference in magnetic susceptibility between sorbitol and sucrose completely separated the two substances in compressed oxygen gas using a gradient MF. In another study, OMF treatment of cantaloupe slowed down respiration and improved shelf life.

Magnetic freezing can produce tiny crystals throughout the body of the frozen product that helps avoid tissue damage of food materials and retain fresh food qualities after thawing. However, such advantages of MF exposure are not consistently observed. MF is assumed to directly affect water molecules by causing vibration, orientation, and/or spinning to prevent the formation of clusters, and thus also facilitate supercooling. The efficacy of MF on frozen food items and its effect during the freezing process is not well understood, and therefore needs more rigorous experimentations.

9.4 Equipment

The MF can be produced when an electric current is passed through a coil of wire, turning it into an electromagnet. Near the coil, MF gets concentrated. The strength of the MF can be enhanced by increasing the number of loops in the coil, increasing the cross-section of each loop, and increasing the current through the coiled wire. Additionally, inserting an iron core into the coil can further improve the production of MF up to 3 T. However, with coil and iron core (ferromagnetic or paramagnetic material), generating MF more than 3 T is difficult due to magnetic saturation in the iron core. To generate MF over 3 T, air-core solenoids are utilized. During the production of MF with intensity as high as 20 T, the temperature rises due to Joule heating. Superconducting magnets, coils capable of producing DC fields, and coils powered by energy discharge stored in a capacitor can generate high intensities of up to 20 T with or without minimum Joule heating (**Fig. 9.4**). Metals can behave

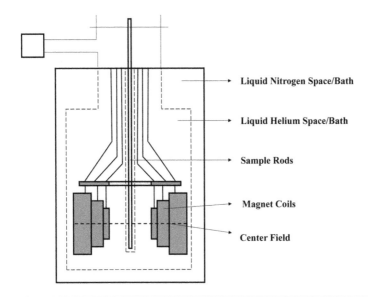

Liquid Nitrogen Space/Bath

Liquid Helium Space/Bath

Sample Rods

Magnet Coils

Center Field

Figure 9.4   Cross-section view of a superconducting magnet. (Adapted from Grigelmo-Miguel et al. [2011])

as superconductors (allows the passage of current with no or negligible resistance) in the presence of liquid Helium, whose temperature can be continually kept low with the help of liquid nitrogen.

MFs with flux densities of more than or equal to 30 T can be produced using hybrid magnets, which combine a superconducting magnetic coil with a water-cooled magnetic coil. The coil stores electricity in the capacitor bank, which is charged from a voltage source. Energy stored in the capacitor bank released in the form of short duration pulses into the coil producing oscillating current, results in the generation of OMF. The polarity of the MF shifts when the current changes direction.

In general, an apparatus for treating food materials with OMF consists of a power source, an AC voltage transformer connected with a capacitor bank, and resistors (together forming a control panel to vary MF strength). A treatment chamber is fitted with an iron core inductor (coil) or superconducting electromagnet or a suitable hybrid magnet (**Fig. 9.5**). The treatment chamber is equipped with a thermocouple and data logger to continuously monitor temperature or a thermostat to monitor and control the chamber temperature. To measure the MF or magnetic flux density (B), one Tesla meter is also connected to the treatment chamber.

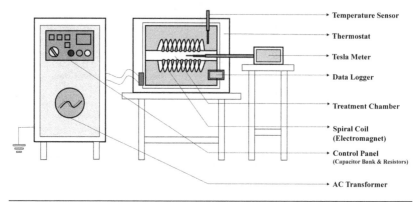

Figure 9.5 Schematic of an oscillating magnetic field apparatus for processing food products. (Adapted from Chitravathi and Chauhan [2019])

The OMF was employed in decaying or fixed amplitude sinusoidal waves. An MF can be homogeneous or heterogeneous. In the homogeneous method, the field amplitude *B* was uniform within the region surrounded by the MF coil. In heterogeneous field *B*, nonuniformity and intensities diminish as one goes out from the center of the coil. OMF is administered in pulses having a reversed charge for each pulse. Over time, the amplitude of each pulse drops progressively to 10% of its initial value.

9.5 Critical Processing Factors

9.5.1 Type of Product

Electrical characteristics greatly influence the efficacy of OMF. The essential criterion for properly preserving food using MF technology is high electric resistivity of more than 10–25 $\Omega \cdot$ cm. Fortunately, many foods exhibit electrical resistivity in this range; for instance, orange has a resistivity of 30–156 $\Omega \cdot$ cm and tomato juice has a resistivity of 40–265 $\Omega \cdot$ cm. The resistivity offered by food materials is further associated with the temperature of the sample or the treatment chamber and the thickness or volume of the product.

9.5.2 Target Microorganism or Enzyme

The potential of OMFs to inactivate microorganisms is widely recognized. Based on the varied effects of MFs on microorganisms, MFs may work under the window effect. This indicates that a specific range of MF can affect the target microorganism and enzyme. MFs, in general, alter the route of metabolism, affect cell organelles and membranes, which may support their growth or inactivate them. Microorganisms may show different susceptibility under different phases of the growth stage (lag, log, stationary, and death phase). The inactivation or inhibition mechanisms for microorganisms and enzymes exposed to OMF are not entirely understood, which sometimes make its effect unclear on the target microorganism or enzyme. However, it can be understood that a higher initial count of microorganisms or concentration of an enzyme can reduce the efficacy of OMF.

9.5.3 Equipment and Mode of Operation

MFs either stimulated or inhibited microbial growth or had no impact in certain situations. The impact of MFs on the microbial population or enzyme activity in the foods varies according to the type of OMF apparatus, MF intensity, number of pulses, frequency, and product characteristics (e.g., resistivity, electrical conductivity, and thickness), and temperature of the treatment chamber (maintained using a thermostat) that can influence the electrical properties of the food material. However, the consequences of these studies are unclear. Exposure to heterogeneous MF can easily diminish the lethality of OMF treatment. Thus, proper homogeneous distribution of MF in the treatment chamber is necessary.

Electromagnets with an alternating current can create OMF. The field strengths are high compared to the Earth's MF, ranging from 5 to 100 Tesla. For a period ranging from a few milliseconds to 25 seconds, frequencies ranging from 5 to 500 kHz are used. This contributes to vegetative cell inactivation. However, if the frequency exceeds 500 kHz, the food becomes heated, making the inactivation process a thermal treatment.

9.6 Practical Example

We will take up one of the research works as a case study. Lipiec et al. (2005) explored the effect of OMF pulses on oat sprouts in terms of total bacterial count (TBC), total fungal count (TFC), total polyphenol content (TPC), and total antioxidant capacity (TAC). Oat grains were sown in petri dish kept under a controlled environment of 21°C, 65% relative humidity, and dark conditions. The grains were allowed to grow into sprouts for 5 days.

OMF treatment: The oscillating and pulsating MF was generated by a solenoid connected with a capacitor bank having a total capacitance of 1.5 mF. The capacitor bank was charged with a high-voltage supply with a maximum output of 5 kV and an electric current of 100 mA. The solenoid consisted of a single-layered coil made from wounding copper wires of a 20–40 mm² rectangular cross-section area. The food samples stored within a test tube were kept inside the solenoid, having an internal diameter of 17 mm during treatment. The oat sprouts were exposed to OMF of 5–8 T for 1–10 pulses. The treatment conditions and the corresponding results have been summarized in **Table 9.1**.

Observations: OMF treatment of oat sprouts helped in reducing the microbial load. At treatment condition of 5×5 T reduced bacterial count by 51% and fungal count by 91% (**Table 9.1**). Subsequently, OMF processing with 1×8 T reduced bacterial count only by 13%, which was far low in comparison to the severity of 5×5 T OMF. Total fungi were seemed to be more susceptible to OMF than total bacteria in oat sprouts. The researchers have attributed the formation of free radicals by OMF and associated damage to cellular protein, fat, DNA, and carbohydrate as the reason for microbial inactivation. On the other hand, phenolic content and antioxidant capacity

Table 9.1 Effect of OMF Pulses on Total Bacterial Count (TBC), Total Fungi Count (TFC), Total Polyphenol Content (TPC), and Total Antioxidant Capacity (TAC) of Oat Sprouts

OMF (T)	NUMBER OF PULSES	TBC (CFU·g⁻¹)	TFC (CFU·g⁻¹)	TPC (mg CATECHIN·(100 g)⁻¹)	TAC (% INHIBITION)
Untreated	-	242	14.6	596.7	36.2
3	10	–	–	896.8	48.0
5	5	119	1.3	–	–
8	1	211	–	582.7	37.6
8	3	–	–	595.0	28.3

Unit T stands for Tesla, and CFU means colony-forming unit. An average control value for the untreated conditions in all cases has been provided, and the rest of the response data has been normalized as per the average control value.

improved at low OMF conditions. Still, in higher OMF conditions, TPC and TAC either remained the same or got reduced by a small degree. The majority of the polyphenols show antioxidant power, and thus TPC and TAC are interrelated. During OMF treatment with a relatively mild condition of 10×3 T, TPC increased by 50%, and similarly, TAC also increased by 33% (**Table 9.1**). A relatively severe OMF processing condition of 1×8 T and 3×8 T, TPC remained almost the same, and the changes were statistically insignificant (p-value < 0.05). TAC showed an insignificant (p-value < 0.05) rise of 4% at 1×8 T but it decreased by 22% after 3×8 T OMF treatment. The sprouts were intact while processing. It is known that plants produce polyphenols and antioxidants as a defensive response to stress or any attack because these components possess antimicrobial antioxidant that prevents oxidative damage and anticarcinogenic properties. Thus, at mild OMF treatment, TPC and TAC increased, but further, it decreased but got balanced by an equivalent generation of natural TPC and TAC by oat sprouts. A significant decrease in TAC at most severe OMF conditions was observed, probably because the oxidative damage of free radicals exceeded the preventive action of the naturally produced TPC and TAC.

9.7 Challenges

Strong SMFs or OMFs (5–50 Tesla) have the potential to inactivate vegetative bacteria. The impulse lasts between 10 milliseconds to several milliseconds. The frequencies are restricted to 500 MHz; beyond this, food temperature rises sharply. Advantageously, in-pack treatment of food is possible with OMF processing. The effects of MFs on microbial organisms have resulted in controversial conclusions. Consistent findings on the method's efficacy are required before adopting this technique for food preservation. OMF treatment affects different food products, microorganisms, enzymes, and associated nutrients. Thus, it has too much variation in its efficacy to act as a food processing and preservation technique. Due to the apparent lack of market acceptance of MFs as an alternative food processing technology, industries want confirmative data on microbial growth and death kinetics.

Furthermore, there is a significant shortage of information about the resistant pathogens, indicator microorganism or enzyme or nutrient for pasteurization/sterilization, microbial inactivation, spoilage enzyme inactivation, associated kinetics, governing mechanism, and critical process factors for the desired goal. Using superconducting technology, the application of MFs might be efficient and advantageous for biomass production at a regulated growth rate. This will aid in the growth of the fermentation and pharmaceutical sectors. Protein and fat structural modifications can be achieved by immersing these constituents in MFs. However, superconductive technology for OMF can be very costly compared to other OMF producing devices.

9.8 Summary

- An alternating or direct current (AC/DC) in conductors or permanent magnets generates an MF.
- In OMF processing, food can be treated packed within plastic or glass containers.
- The magnetic flux density, frequency, waveform, exposure duration (time), and polarity are the characteristics evaluated for the categorization of OMFs.
- Microbial inactivation can occur due to ion translocation under MF via ICR or IPR mechanisms and DNA damage by free radical or transferring coupling energy.
- Higher energy (flux density of 5–10 T and frequencies of 5–500 kHz) OMFs may be utilized in food processing.
- The impact of OMF on the microbial population of foods varies according to MF intensity, number of pulses, frequency, and product characteristics (e.g., resistivity, electrical conductivity, isotropy, and magnetic susceptibility).
- For producing higher intensity OMF ≤ 3 T, an iron core is inserted inside the coil, whereas for producing OMF > 3 T, air-core solenoids are used.
- If the frequency exceeds 500 kHz, the food becomes heated, causing thermal degradations.
- Superconducting magnets can generate OMF up to 20 T without Joule heating.

- MFs with flux densities of more than or equal to 30 T can be produced using hybrid magnets, which combine a superconducting magnetic coil with a water-cooled magnetic coil.

9.9 Solved Numerical

1) An electromagnet with AC generates an OMF, whose intensity is controlled by the current (I) flowing in the coil. The electrical current flow through winding coils produces heat because of the resistance offered by the coils. This heating effect is known as 'Joule heating', 'Resistivity heating', or 'Ohmic heating'. Using the following information, calculate the overall resistance (Z) offered because of the coil resistance (R) and inductive reactance (X_L). Also, find the heat generated due to Joule heating if the operating voltage is 30 V and processing time is 10 min. Given information: $R = 7\,\Omega$; $X_L = 3\,\Omega$

Solution: Use extended Ohm's law to estimate the peak current (I) and overall resistance (Z).

Extended Ohm's law states,

$$I = \frac{V}{|Z|} = \frac{V}{\sqrt{R^2 + X_L^2}}$$

$$|Z| = \sqrt{R^2 + X_L^2} = \sqrt{7^2 + 3^2} = 7.62\,\Omega$$

To find out the peak current (I),

$$I = \frac{V}{\sqrt{R^2 + X_L^2}} = \frac{30}{7.62} = 3.94\text{A}$$

To find the heat generated due to Joule heating, we can use the following equation, *Heat* (W) $= V \cdot I \cdot t = 30 \times 3.94 \times (10 \times 60) = 70920\text{J} = 70.92$ kJ

2) An electromagnet with AC generates an OMF using a coil. Calculate the magnetic flux density (B) and magnetic field intensity (H) if the current (I) flowing through a length (L) of a 2 m long coil having 100 turns (N) is 5 A.

Solution: Use Ampere's law to estimate the magnetic flux density (B) and field intensity (H).

Using Ampere's law, we have

$$B = \mu_0 \times \frac{N}{L} \times I = 4\pi \times 10^{-7} \times \frac{100}{2} \times 5 = 3.1416 \times 10^{-4} T = 0.0314 \ \mu T$$

Now, from the basics of electromagnetism, we know that $B = \mu_0 H$

Using the above two equations, we can calculate magnetic field intensity (H) that,

$$H = \frac{B}{\mu_0} = \frac{N}{L} \times I = \frac{100}{2} \times 5 = 250 \ A \cdot m^{-1}$$

3) A study reported the viable cell counts of three different pathogens in a 1 cm^3 treatment chamber when the samples are treated with an OMF with varying magnetic flux density (B). The processing time was kept at 10 min and 30 min at $B = 5$ T, and the microbial count is tabulated below. Find the rate constant and half-life for each pathogen. Further, comment on their resistance to OMF processing.

	PATHOGEN (CFU·mL⁻¹)					
MAGNETIC FLUX DENSITY (B)	Ervinia carotovora		Alternaria solani		Streptomyces scabies	
Control (B = 0 T)	10.3 × 10⁸		9.7 × 10⁶		9.7 × 10⁶	
Processing Time →	10 min	30 min	10 min	30 min	10 min	30 min
Treated (B = 5 T)	8.1 × 10⁸	4.8 × 10⁷	9.3 × 10⁶	1.1 × 10⁶	4.6 × 10⁶	9.4 × 10⁴

Given information:

- Assume the initial microbial cell counts as the same data given for control (0 T).
- Assume that the microbial destruction kinetics follows the first-order kinetics model.

Solution: The microbial count of control (0 T) is taken as the initial count. So, $(N_0)_{Ec}$, $(N_0)_{As}$, and $(N_0)_{Ss}$ are 10.3×10^8, 9.7×10^6, and 9.7×10^6, respectively.

a. Using first-order kinetics, we know that,

$$\ln\left(\frac{N}{N_0}\right) = -kt$$

For *E. carotovora* (putting the initials in subscript for referring to the specific k; k_{Ec}),

$$\ln\left(\frac{8.1\times10^8}{10.3\times10^8}\right)=-k_1\times10; k_1=0.024\,\text{min}^{-1}$$

$$\ln\left(\frac{4.7\times10^7}{10.3\times10^8}\right)=-k_2\times30; k_2=0.103\,\text{min}^{-1}$$

$$k_{Ec}=\frac{0.024+0.103}{2}=0.0635\,\text{min}^{-1}$$

Similarly, for *A. solani* and *S. scabies*, we can calculate,

$$k_{As}=\frac{k_3+k_4}{2}=\frac{\left\{\ln\left(\frac{9.3\times10^6}{9.7\times10^6}\right)\div10\right\}+\left\{\ln\left(\frac{1.1\times10^6}{9.7\times10^6}\right)\div30\right\}}{2}$$

$$=\frac{0.004+0.072}{2}=0.038\,\text{min}^{-1}$$

$$k_{Ss}=\frac{k_5+k_6}{2}=\frac{\left\{\ln\left(\frac{4.6\times10^6}{9.7\times10^6}\right)\div10\right\}+\left\{\ln\left(\frac{9.4\times10^4}{9.7\times10^6}\right)\div30\right\}}{2}$$

$$=\frac{0.075+0.154}{2}=0.1145\,\text{min}^{-1}$$

Using the half-life ($t_{1/2}$) formula based on k-value, we have $t_{1/2}=\frac{0.693}{k}$ So $t_{1/2}$ (in minutes) for each microbe is as follows:

$$\left(t_{1/2}\right)_{Ec}=\frac{0.693}{k_{Ec}}=\frac{0.693}{0.0635}=10.91\,\text{min}$$

$$\left(t_{1/2}\right)_{As}=\frac{0.693}{k_{As}}=\frac{0.693}{0.038}=18.24\,\text{min}$$

$$\left(t_{1/2}\right)_{Ss}=\frac{0.693}{k_{Ss}}=\frac{0.693}{0.1145}=6.05\,\text{min}$$

We can conclude that the higher the k value, the easier it is to kill the microbe, or the lower the k value, the more resistant the microbe is.

So, for the OMF processing condition of 5 T, we can say that *A. solani* is more resistant, followed by *E. carotovora* and *S. scabies*. Based on the $t_{1/2}$ values also, we can confirm this sequence.

9.10 Multiple Choice Questions

1. The ratio of the amplitude of the MF to the amplitude of the electric field for electromagnetic wave propagation in a vacuum is equal to:
   a. Unity
   b. Speed of light in vacuum
   c. Reciprocal of the speed of light in vacuum
   d. The ratio of magnetic permeability to electrical suscepti-bility in a vacuum.

2. If a charged particle moves through a magnetic field perpen-dicular to it, then:
   a. Both momentum and energy of particle change
   b. Energy remains constant, but momentum changes
   c. Momentum, as well as energy, remain constant
   d. Momentum remains constant, but energy changes

3. When an electron is projected along the axis of a circular con-ductor carrying the same current, it will experience:
   a. No force
   b. A force perpendicular to the axis
   c. A force at an angle of 4° with the axis
   d. A force along the axis

4. If an electron is moving with velocity ν producing a magnetic field B, then:
   a. The direction of field B will be the same as the direction of velocity ν
   b. The direction of field B will be opposite to the direction of velocity ν
   c. The direction of field B will be perpendicular to the direction of velocity ν
   d. The direction of field B does not depend upon the direction of velocity ν

5. In electromagnetic waves, the phase difference between electric and magnetic field vectors are:
   a. zero
   b. $\pi/4$
   c. $\pi/2$
   d. π

6. The oscillating magnetic flux density in a plane electromagnetic wave is given as; $B = (8 \times 10^{-6}) \cdot \sin \{2 \times 10^{11} \cdot t + 300\pi \cdot x\}$ T. The wavelength (λ) of the EM wave is _____.
   [Hint: $B = B_0 \cdot \sin\{\omega \cdot t + k \cdot x\}$, and $k = 2\pi/\lambda$].
   a. 0.80 cm
   b. 0.67 cm
   c. 2×10^{-2} cm
   d. 1×10^3 m

7. A charged particle oscillates about its mean equilibrium position with a frequency of 10^9 Hz. The frequency of electromagnetic waves produced by the oscillator is:
   a. 2×10^9 Hz
   b. 3.16×10^4 Hz
   c. 5×10^8 Hz
   d. 10^9 Hz

8. If speed of gamma rays, X-rays, and microwave are v_g, v_x and v_m, respectively, then:
   a. $v_g > v_x > v_m$
   b. $v_g < v_x < v_m$
   c. $v_g > v_m > v_x$
   d. $v_g = v_x = v_m$

9. A system/process where input is converted into an output without any knowledge of the governing mechanism is called as _____.
   a. Schrödinger system
   b. Black-box system
   c. White-box model/system
   d. Glass-box model/system

10. An ion having charge q enters a magnetic field B at a velocity v and experiences a force F. The angle between v and B is θ. Then the correct relationship is:

a. $F = q \cdot v \cdot B \cdot \tan\theta$
b. $F = q \cdot v \cdot B \cdot \cos\theta$
c. $F = q \cdot v \cdot B \cdot \sin\theta$
d. $F = q \cdot B/v$

9.11 Short Answer Type Questions

a. What are the effects of an OMF on pathogens in food?
b. What is magnetic freezing, and how can it be created using an OMF?
c. How are static, oscillating, or pulsed magnetic fields different from each other?
d. How does an OMF is generated?
e. Highlight the theories explaining the MF-based inactivation of spoilage enzymes in food.

9.12 Descriptive Questions

a. Explain the various applications of an OMF in food preservation.
b. Describe the critical process factors influencing the effectiveness of OMF treatments on microbial populations.
c. Elaborate on the two primary mechanisms/theories of microbial inactivation using an MF in a food product.
d. What is the physical significance of magnetic flux density? How charged particle behaves in an MF?
e. Define Joule heating. With a schematic of a superconducting magnet-based OMF system, discuss how superconductors can affect the joule heating phenomenon.

9.13 Numerical Problems

1) A study reported the viable cell counts of three different pathogens in a 1 cm^3 treatment chamber when the samples were treated with an OMF with varying magnetic flux density (B). The processing time was kept

at 10 min for each process, and the microbial count is tabulated below. Assuming a first-order kinetic model, find each pathogen's rate constant and half-life. Further, comment on their resistance to OMF processing.

MAGNETIC FLUX DENSITY (B)	PATHOGEN (CFU·mL^{-1})		
	Ervinia carotovora	*Alternaria solani*	*Streptomyces scabies*
Control (0 T)	10.3×10^8	9.7×10^6	9.7×10^6
5 T	8.1×10^8	9.3×10^6	4.6×10^6
20 T	2.5×10^5	3.1×10^5	9.7×10^3

2) A study was carried out to understand the effect of OMF on freezing time and supercooling of deionized water. The freezing time can be calculated using Plank's equation as given here:

$$t_f = \frac{\rho L}{T_F - T_M} \left[Pa \left(\frac{1}{h} + \frac{b}{k_p} \right) + \frac{Ra^2}{k} \right]$$

Where t_f is the freezing time (s), ρ is the density of sample (1000 kg/m^3), L is the latent heat of fusion (334 kJ/kg), h is the convective heat transfer coefficient (1 kW/m^{2}K), a is the sample thickness (1 cm), b is the container thickness (1 mm), k_p is the thermal conductivity of container (0.3 W/mK), k is the thermal conductivity of sample (20 mW/mK), T_F is the freezing point of the sample (273 K), T_M is the medium temperature (263 K), and P and R are the shape factors. The shape factors for the infinite slab used are 1/2 and 1/8, respectively. Using the above data, calculate the freezing time.

References

Lipiec, J., Janas, P., Barabasz, W., Pysz, M., & Pisulewski, P. (2005). Effects of oscillating magnetic field pulses on selected oat sprouts used for food purposes. *Acta Agrophysica*, 5(2), 357–365.

Suggested Readings

Barbosa-Canovas, G. V., Schaffner, D. W., Pierson, M. D., & Zhang, Q. H. (2000). Oscillating magnetic fields. *Journal of Food Science*, 65, 86–89.
Miñano, H. L. A., de Sousa Silva, C. A., Souto, Á., & Costa, E. J. X. (2020). Magnetic fields in food processing perspectives, applications and action models. *Processes*, 8(814), 1–11.

Chitravathi, K., & Chauhan, O. P. (2019). Pulsed Magnetic Field Processing of Foods. In O. P. Chauhan (Ed.), *Non-thermal Processing of Foods*. CRC Press, Taylor and Francis Group, Boca Raton, Florida, USA. pp. 261–282.

Grigelmo-Miguel, N., Soliva-fortuny, R., Barbosa-Cánovas, G. V., & Martín-Belloso, O. (2011). Use of Oscillating Magnetic Fields in Food Preservation. In H. Q. Zhang, G. V. Barbosa-Cánovas, V. B. Balasubramaniam, C. P. Dunne, D. F. Farkas, & J. T. Yuan, (Eds.), *Nonthermal Processing Technologies for Food*. Wiley-Blackwell, Chichester, West Sussex, UK. pp. 222–235.

Answers for MCQs (sec. 9.10)

1	2	3	4	5	6	7	8	9	10
c	b	a	c	a	b	d	d	b	c

10

COMMERCIAL ASPECTS
AND CHALLENGES

10.1 Potential and Limitations

Non-thermal techniques, such as high-pressure processing (HPP), pulsed electric field (PEF), ultraviolet (UV), pulsed light (PL), power ultrasound (US), cold plasma (CP), ozone (O_3), oscillating magnetic field (OMF), and irradiation (ionizing radiation), have proved their potential to overcome shortcomings of thermal processing and satisfy the demands of all the stakeholder including consumers, sellers and retailers, and industries. However, these techniques have limitations in cost, selective lethality, product suitability, etc. **Table 10.1** summarizes the advantages and limitations of the non-thermal technologies for food preservation.

10.2 Sustainability

Sustainability means managing the resources without compromising the needs of future generations. The three pillars of sustainability are economy, environment, and society (**Fig. 10.1**). The environmental aspect of sustainability emphasizes the process design to preserve the utilization of natural resources. To maintain environmental sustainability, the economic sustainability of a process comes into the picture. Using renewable energy and efficiently using the same helps in economic sustainability. It helps preserve the environment and helps reduce expenditure, extending the life cycle of the process. The effect of process operations on the people accounts for sustainability's social aspects. Some strategies to make a non-thermal technique sustainable are as follows:

- Increasing energy efficiency of the non-thermal processing
- Reusing of energy going to waste
- Processing of local foods and value addition

DOI: 10.1201/9781003199809-10   **189**

Table 10.1 Advantages and Limitations of Non-thermal Food Preservation Technologies

TECHNOLOGY	ADVANTAGES AND LIMITATIONS
HPP	**Advantages**
	a. Uniform distribution of pressure takes place
	b. Allows food processing at ambient or even lower temperatures
	c. Can inactivate most of the pathogens, protozoa, parasites, and spoilage microorganisms, such as yeast-mold, Gram-positive, and Gram-negative bacteria
	d. It can inactivate spoilage enzymes and also help in retarding enzymatic-browning
	e. Minimal impact on food quality involving sensory and nutrient content because HPP does not break covalent bonds
	f. Unlike thermal, the effect of high pressure is not dependent on the size and shape of the food material
	g. The temperature rises during the compression, and it goes back to the initial temperature after decompression (adiabatic heat of compression is reversible)
	h. In-pack treatment is possible with HPP
	i. Digestibility of certain food and bioavailability of nutrients can be enhanced
	Limitations
	a. Yeasts, molds, and bacterial spores are highly resistant to HPP
	b. Sometimes the pressure-assisted enzyme inactivation can be a reversible process
	c. Some food products are not suitable for HPP; solid food having air inside (like bread), food packed in rigid-type packaging material (like glass), and food having low moisture content (like spices, dry fruits) usually below 40%
	d. The capital cost of HPP equipment is very high
	e. Equipment handling is complex and requires significant precision, and therefore having a separate expert for operating HPP is important
	f. Takes extra time in compression and decompression thus increases the overall process time and makes it difficult to modify HPP into a continuous process
PEF	**Advantages**
	a. Can effectively inactivate a wide range of microorganisms
	b. It has the potential to inactivate spoilage enzymes
	c. Allows processing at ambient pressure and with negligible temperature rise
	d. Minimal impact on food quality involving sensory and nutrient content
	e. Have great perspectives for energy saving while processing
	f. Usually requires very short treatment time
	Limitations
	a. Not suitable for food products either with poor capacitance and dielectric properties or with high conductivity (i.e., low resistivity)
	b. Non-fluid type food products can only be treated in the batch mode provided the dielectric breakdown is avoided
	c. High equipment cost
	d. Air bubbles cannot withstand strong electric fields, and thus solid food containing air bubbles is not suitable for PEF processing
	e. Bacterial spores and fungal spores are highly resistant toward PEF
	f. The effect on enzymes is not properly explored yet
	g. Possibility of occurrence of electrochemical reaction between electrode and food sample (based on equipment design) that can reduce the electrode life and make the food toxic

Table 10.1 *(Continued)* Advantages and Limitations of Non-thermal Food Preservation Technologies

TECHNOLOGY	ADVANTAGES AND LIMITATIONS

UV & PL

Advantages

a. PL operates on nontoxic noble gases like xenon gas
b. The majority of the microorganisms can be inactivated using UV and PL
c. UV processing has a simple set-up, convenient handling, and low equipment cost
d. UV and PL can also be used to disinfect food contact surfaces
e. UV and PL present low operation cost
f. PL can reach adequate lethality even in a short treatment time
g. Both UV and PL can show significant good energy efficiency; however, the energy consumption of UV is smaller than PL, but at the same time PL based inactivation is faster than UV
h. Sometimes UV and PL treatment may enhance phenolic content and antioxidant capacity of the sample

Limitations

a. Fungal spores, bacterial spores, and some spoilage enzymes are highly resistant to UV and PL
b. UV has lower penetration power than PL
c. UV or PL cannot penetrate solid food, semi-solid food having poor light transmittance, and liquid food that are optically opaque, translucent, or possess a large number of suspended particles
d. Longer processing time with PL can increase product temperature
e. PL treatment may initiate the formation of ozone in the presence of oxygen
f. In some cases, UV and PL can degrade natural pigments, nutrients, and affect sensory characteristics
g. PL imposes a high-initial investment cost

Power ultrasound

Advantages

a. Can inactivate microorganisms with less or negligible effect on food quality and sensory attributes
b. Unlike few other non-thermal technologies where sub-lethally damaged cells can heal themselves under favorable conditions, cells ruptured or disintegrated by ultrasound cannot recover
c. Sonication of solid foods like meat can help in improving tenderness, cohesiveness, and water-binding capacity
d. It has many other food applications besides processing and preservation, such as extraction, drying, dehydration, filtration, freezing, degassing, and defoaming
e. The equipment cost is relatively low

Limitations

a. Longer treatment can increase sample temperature and loss of nutrients
b. Overall lethality of power ultrasound is relatively lower than other non-thermal techniques, such as HPP, PEF, UV, and PL
c. Spores, fungi, and spoilage enzymes are resistant to power ultrasound and thus require longer processing time
d. The presence of air or air bubbles in the food sample diminishes the lethal impact of ultrasound
e. Sonication is more suitable for fluid food, even though solid foods can be treated in bath mode, but the sound waves can't penetrate deeper
f. In the case of probe sonicators, there is a chance of food contamination from the corrosion of the metallic horn

(Continued)

Table 10.1 (*Continued*) Advantages and Limitations of Non-thermal Food Preservation Technologies

TECHNOLOGY	ADVANTAGES AND LIMITATIONS

Cold plasma

Advantages

a. Capable of inactivating microorganisms and spoilage enzymes with adequate lethality
b. Can degrade food allergens, pesticides, and toxins
c. The cost of equipment is low compared to HPP, PEF, and irradiation set-up
d. Liquid food can be processed by releasing plasma directly in the liquid phase, and cold plasma can show its lethality through plasma-activated water
e. It can act as an excellent surface disinfectant
f. There exist many different types of cold plasma instruments, including both food contact and noncontact type
g. In some cases of solid food treatment with cold plasma may cause improved tenderness, water uptake, and solubility in water

Limitations

a. It cannot penetrate solid or semi-solid food and can act only on the surface
b. Spores, fungi, and to some extent, yeast also show significant resistance against cold plasma
c. The complexity of instruments which requires trained personnel and precision
d. Mechanisms and effects of cold plasma processing on microbes, enzymes, nutrients, and sensory properties are unexplored
e. Plasma may cause lipid oxidation, and contact type cold plasma systems like dielectric barrier discharge (DBD), may contaminate the food sample
f. Certain devices like plasma jet or microwave powered plasma, sample temperature may exceed 60°C
g. Use of costly noble gases as a feed gas for plasma production and installation of safety system regarding the high voltage supplier in cold plasma system can add up to the equipment cost

Ozone processing

Advantages

a. Can inactivate a wide spectrum of microorganisms, including: bacteria, bacterial spore, yeast, fungi, fungal spore, virus, and protozoa
b. Ozone kills microorganisms through cell lysis, which prevents the microbes from developing resistant strains
c. Ozone can deteriorate pesticides, mycotoxins, and unpleasant odor-causing volatiles generated from waste material
d. Relatively low equipment cost and short processing time
e. Ozone exposure can be delivered to food products both in gaseous form and in the liquid phase (ozone activated water)
f. Acts as an excellent surface disinfectant
g. The remaining ozone breaks down to O_2 after processing, leaving no toxic residue

Limitations

a. Food rich in lipid is not suitable for ozone treatment
b. Prolonged exposure to ozone can affect sensory, surface morphology, and can also lead to sample discoloration
c. Ozone works better on the surface and cannot penetrate deeper into the product, especially in solids
d. To prevent quick ozone decay in the aqueous phase, the liquid medium should have pH < 7 and sample temperature preferably < 35°C

Table 10.1 *(Continued)* Advantages and Limitations of Non-thermal Food Preservation Technologies

TECHNOLOGY	ADVANTAGES AND LIMITATIONS
Ozone processing *(Continued)*	e. The presence of certain organic components may interfere with the action of ozone by competing for O_3 along with microorganisms f. The effect of ozone on spoilage enzymes is not much explored g. Ozone must be produced on the spot of its usage because of its unstable nature h. Ozone can erode the surface of metal food containers, and its inhalation is harmful to humans
Irradiation (ionizing radiation)	**Advantages** a. Effective against many foodborne pathogens b. Microorganisms are attacked in two ways: by *directly* damaging cell's genetic materials and organelles and by *indirectly* attacking with radicals, hydrogen peroxide, and ozone formed via radiolysis of available water c. Do not cause any major effect on the nutritional content of food, with little effect on sensory properties, and treated products are safe to consume d. In-pack treatment is possible with irradiation e. Fungal toxins are degradable by γ-irradiation f. Can induce depolymerization of starch and pectin, leading to softening **Limitations** a. Costly and complex irradiation set-up or treatment plant b. Adequate safety and continuous maintenance is required because ionizing radiation (especially γ-rays) is lethal for humans, and radiation-based technology using radioactive elements is always associated with national security c. Lethality of irradiation will reduce in the case of food with low water activity d. Microbial toxins, viruses, spores, and spoilage enzymes are resistant to degradation or inactivation e. Bacterial species of *Deinococcus* are highly resistant to ionizing radiation due to their highly efficient DNA repair system f. May show detrimental effects on food quality, such as oxidation of lipids, loss of antioxidants, and oxidation of useful proteins
OMF	**Advantages** a. OMF has the potential to inactivate microorganisms b. In-pack treatment is possible with OMF c. Applying OMF during freezing can help in producing tiny crystals volumetrically and thus avoiding tissue damage of food materials and also facilitating super cooling d. The instrumentation cost seems less e. Superconducting magnet allows the production of MF above 3 T with minimum or without Joule heating **Limitations** a. The effect of OMF on every microorganism, enzyme, or food component is unique and making the effect of OMF highly uncertain and inconsistent b. Spores and spoilage enzymes are highly resistant to OMF c. Mechanisms for the effect of OMF on microbes, enzymes, and nutrients are not completely understood d. Microbial inactivation can be seen only for food material having an electrical resistivity of more than 10–$25\ \Omega\cdot cm$ e. During the production of MF with intensity as high as 20 T, temperature rises due to Joule heating occurs

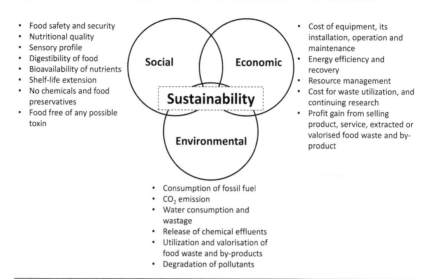

- Food safety and security
- Nutritional quality
- Sensory profile
- Digestibility of food
- Bioavailability of nutrients
- Shelf-life extension
- No chemicals and food preservatives
- Food free of any possible toxin

Social

Economic

Sustainability

Environmental

- Cost of equipment, its installation, operation and maintenance
- Energy efficiency and recovery
- Resource management
- Cost for waste utilization, and continuing research
- Profit gain from selling product, service, extracted or valorised food waste and by-product

- Consumption of fossil fuel
- CO_2 emission
- Water consumption and wastage
- Release of chemical effluents
- Utilization and valorisation of food waste and by-products
- Degradation of pollutants

Figure 10.1   Sustainability aspects for non-thermal technology in food processing sector.

- Formulation of value-added product from the waste using the technique
- Reducing water usage during processing
- Designing the process considering safety, stability, and quality of product
- Coming up with consumer-driven innovations
- Going for the circular economy within the cycle

10.2.1 Social Impacts

As non-thermal technology involves lower temperatures than thermal technology, the nutritional and sensory quality of the product is generally maintained. HPP may enhance digestibility of food and has been proven to increase the bioavailability of micronutrients and phytochemicals, reduce allergens, and preserve healthy lipids. The retention of thermosensitive compounds in the food product has been the key aspect for all the non-thermal techniques. In this sense, these techniques fulfill consumer demand for healthy and nutritious food. However, there are certain examples of non-thermal processed causing undesirable damage to the quality of the food product. Irradiation has been seen to cause slight changes in the flavor, nutrients by inducing changes in the bioactive compounds.

Similarly, PEF has also been seen to modify phenolic compounds, vitamin C, and antioxidants. Ozone treatment has higher effects on the antioxidant compounds due to its oxidative nature. It may also activate oxidizing enzymes, such as ascorbate oxidase, causing ascorbic acid degradation.

The microbial safety aspect of non-thermally processed food is well established. All the non-thermal techniques are very effective against foodborne pathogens. Many non-thermal technologies (ozone, cold plasma, and irradiation) can degrade toxins (like mycotoxins), food allergens, and pesticides. There is evidence that spore formers may not be fully inactivated by non-thermal stress. However, the safety of the food is not being compromised in any case. Moreover, the pricing of the non-thermally treated product limits their application for low-value high-bulk products. Overall, for society, non-thermal technologies can provide microbiologically safe products with higher nutrient retention than thermal technology without chemical preservatives.

10.2.2 Environmental Impacts

Food processing technologies utilize energy sources primarily from fossil fuels and few from renewable energy sources for sterilization, pasteurization, drying, etc. On the other hand, certain non-thermal technologies reduce fossil fuel usage by reducing energy consumption through cooking or processing time and treatment temperature. Consequently, CO_2 emission is also reduced as heat requirement is negligible. Large steam production for heat transfer processes or fire from burning wood and coal for direct heat application is avoided. Non-thermal technologies potentially reduce energy and water consumption by increasing the efficiency of the process. For example, in PEF, applying mild heat can achieve enzyme inactivation and provide a longer shelf life without refrigeration. PEF pasteurization has proven to utilize less energy than the conventional pasteurization method. Negligible amounts of chemicals are used to be released into the environment. Generally, non-thermal technologies are waste-free, reduce emissions, and utilize fewer amounts of water.

Additionally, the by-products of processing can be valorized using non-thermal technologies. For instance, US, high pressure, and electric field have been used to extract bioactive from by-products. PEF has been used for anthocyanin extraction from purple-fleshed pota-toes, while high pressure has been used to remove the same from elderberry.

Non-thermal assisted bioactive extraction from food waste is also done in industries that help avoid the use of chemicals and solvents, which solve the problems related to the drainage of environmentally harmful chemicals and solvents. Ozone and cold plasma treatment are also employed to degrade odor-causing wasted and pollutants, respectively. Overall, non-thermal technologies are more environ-mentally sustainable than thermal processing due to more efficient energy consumption, renewable resources, reduced carbon emissions, and waste utilization capabilities.

10.2.3 Economic Aspects

Economic sustainability focuses on technology's cost, investment, and profit aspects. It can be further understood through the cost of equip-ment, installation, operation, maintenance, waste reduction or utili-zation, and recycling. The operational costs are associated with the energy requirements of the process. The food industry comes under energy-intensive industry among paper and pulp, chemical, iron, steel, etc., consuming around 4% of energy worldwide. As part of the food industry, food processing consumes one-third of the energy. Internal energy, applied energy, and consumed energy are considered for the energy evaluation. Non-thermal technology mainly consumes electri-cal energy, but internal energy is generated that achieves microbial inactivation. This reduces the requirement of energy input for heat gen-eration during the process. For example, adiabatic and resistive heat is generated in HPP and PEF, respectively. The economic sustainability of HPP depends on the total cycle time involving pressure, energy, labor, and capital. It is an expensive process where the cost has been reported to be seven-times higher than a thermal process required for juice pasteurization. During the HPP, energy conversion from elec-trical to mechanical pressure needs to be efficient. Similarly, for PL and UV, electrical energy is converted into light energy; for OMF, electric energy is converted into a magnetic field; and for γ-irradiation atomic energy is converted into electromagnetic energy possessed by photons. The converted form of energy is specific to the non-thermal technique employed. UV and PL have lower operational costs as compared to HPP and PEF.

Optimization of processing conditions and equipment design can reduce the energy and resource requirements (such as oxygen gas supply for ozone or fresh electrode requirement for PEF after exhausting the previous one) of the process and hence the cost, for instance, by changing the pulse conditions and temperatures in PEF and increasing the electrode lifespan. Using the energy efficiently and recovering the dissipated energy to divert its utilization in other processes, such as mild preheating of the food product, would lead to sustainability. Hurdling of technologies (combining technologies), such as UV with PEF or mild heat with HPP, can also reduce the energy requirement. The higher equipment cost is an issue, which is compensated by the market demand of the non-thermal technology. For instance, HPP is commercially successful worldwide, providing more than 150,000 tons of production per year.

Additionally, selling the products extracted or developed by valorizing food wastes or by-products using non-thermal technologies can also add up to the economic gains of the industry. US, HPP, and PEF have been used to extract bioactive from by-products. Overall, the non-thermal technology can be sustainable, which may vary depending on the food sample.

10.3 Commercialization Aspects

10.3.1 Consumer Response

Consumers who are majorly comprised of health-conscious individuals want to have food products that are chemical-free and that have adequate safety, stability, and nutrients that are on par with natural food products. Demand for a sensory and nutritional profile similar to fresh and natural food items has created a demand for *minimally* or *gently* processed food products. Non-thermal pasteurization, sterilization, and other forms of non-thermal processing are being observed as alternatives to conventional thermal technology and fulfill consumer demands. Even after complying with consumer demand, delivering good organoleptic and nutritional products may not guarantee its success. This may happen because, in addition to the good quality ingredients, consumers are also concerned about processes undertaken during the 'farm to fork' food chain and other factors, such as price, origin, and animal welfare. For example, despite having significant advantages compared to thermally treated food, irradiated food could not gain commercial success yet. The majority of the studies reveal that consumers are not familiar with irradiation technology and show fear of possible radiation residues, fear for the presence of carcinogenic components, a possible risk to factory operators, and harm to the environment. Such mis-believes and lack of awareness among the consumers has hampered the motivation of the food industry's implementation for irradiation. However, a recent study shows that replacing the name 'irradiation' with 'ionizing energy' mentioned on the label has improved its acceptance.

High pressure and PEF processed products have received positive responses from the consumers because of their high nutritional and sensory qualities close to their respective natural ingredients. The environment-friendly characteristics of HPP and PEF have also played a significant role in earning the consumer's favor and raising their commercial marketability. PEF is still somewhat new to people; due to their lack of knowledge, they perceived PEF as a process, which involves the usage of electricity for food processing that may bring unknown changes to the food item. Cold plasma-treated products are yet to be tested in the real market, and the majority of the sensory data existing is at the research level only. PL has been commercially exploited for surface disinfection for food contact surfaces and food packages. Ozone and UV treatments are popularly being used for the water disinfection processes. The application of OMF is limited compared to other non-thermal techniques. Therefore, in addition to the knowledge of safety and quality, spreading public awareness about alternative processing technology is essential for positive consumer response and commercial success.

10.3.2 Market Status

The estimated global market of non-thermal processing units and instrumentation was about USD 1,032.34 million in 2020, and it is expected to reach USD 1,889.10 million by 2027 (MarketsandMarkets, 2022). Non-thermal technology is provided by major players, such as Bühler (Switzerland), Hiperbaric España (Spain), CHIC FreshTech (USA), Bosch (Germany), Avure Technologies (USA), Nordion (Canada), Symbios Technologies (USA), Pulsemaster (Netherland),

and so on. Non-thermal processed food products are available in the markets of several countries in the world, such as the countries of the American continents (USA, Mexico, Canada, Argentina, and Brazil), European continent (France, Germany, Italy, Netherland, UK, and Spain), Asia-Pacific and African continents (Russia, China, India, Japan, Saudi Arabia, United Arab Emirates, Singapore, Indonesia, Malaysia, South Korea, Philippines, Australia, and South Africa). North America region leading the market of non-thermal processed food products in 2019. HPP, PEF, and irradiation are the major non-thermal technologies whose food products are currently available in the market. HPP-treated food products dominate the non-thermal processed food market. In addition to that, the meat and seafood segment dominates the non-thermal treated food categories.

10.4 Packaging Challenges

Adequate and compatible packaging material selection for non-thermal processed food products is essential. Non-thermal methods allow working with low-temperature conditions, and thus mandatory requirement of packaging materials having a high-melting point is not compulsory in such cases. Cold sealing techniques, such as adhesives, sealants, and polymers, can be utilized. In HPP, the packaging materials need suitable flexibility, the ability to sustain their physical integrity under high pressure and high oxygen barrier properties. To protect the packaging material from deformation strains, headspace in the package should be kept to a minimum. Damage to the package during HPP happens because air present in the headspace is compressible, and under high pressure, it causes nonuniform pressure distribution on the package. Glass, metal containers, and paperboard-based packaging materials are not suitable for HPP. Packaging materials made of ethylene vinyl alcohol (EVOH) and polyvinyl alcohol (PVOH) are compatible with HPP. Polyethylene (PE) pouches, polypropylene (PP), polyester tubes, nylon PP pouches, laminated plastic films, oriented polypropylene (OPP), cast PP, and polyethylene terephthalate (PET) also show good potential and compatible characteristics.

Packaging of PEF-treated food is usually done after treatment. The product's shelf life primarily depends on the oxygen and water vapor barrier properties. Some of the research studies show that thermoformed plastic containers, glass bottles, PET, high-density polyethylene (HDPE), a combination of high impact polystyrene (HIPS), polyvinylidene chloride (PVDC), and low-density polyethylene (LDPE) have been successfully tested for PEF treated juices.

The nature of irradiation technology allows in-pack treatment, which helps avoid post-treatment recontamination and cross-contamination problems. However, depending on the dose and polymer composition of the package, irradiation can affect flexible type packaging materials. Ordinary glass gets discolored after irradiation exposure. Studies indicate that PE, polystyrene (PS), and PET can be used for irradiated food packaging. Radiation doses up to 30 kGy do not show any significant impact on water vapor permeability and gas permeability (CO_2 and O_2) of ethylene-vinyl acetate (EVA), LDPE, HDPE, biaxial oriented polypropylene (BOPP), and PS films. Some research also reported that irradiation (≥ 25 kGy) of the flexible type packaging material could form volatiles in materials, such as PE, PET, OPP, LDPE, and PP. These volatiles and nonvolatiles produced due to irradiation can affect sensory properties, contaminate the food, and degrade the packaging material.

Achieving in-pack disinfection or sterilization with PL treatment requires packaging material to be transparent to UV radiation and possesses adequate heat resistance. Materials having good heat resistance like PVC can be suitable. However, adequate transmittance of UV is also necessary to attain microbial inactivation in the food sample. Opaque materials like metals and paperboards are not suitable; only UV transparent glass and special polymer (plastic) can be compatible. Ordinary glass blocks the majority of UV radiation. Finally, consumer acceptance of non-thermal processed food products is also affected by the packaging material and delivery information. Information about the novel processing technology should be provided on the package regarding how it works and its key benefits compared to conventional thermal technology.

10.5 Regulatory and Legislative Issues

To commercialize any new product or process, it is necessary to get the approval from the respective regulatory body. The regulatory body aspects of non-thermal processing also have to follow the same legislative process to get approval. Specific regulations for various non-thermal processing methods and products are pretty rare. Thus, in 1997, EU regulation for novel foods and ingredients (CE258/97) established an evaluation, regulation, and a licensing system mandatory for all new foods and processes. The Food Safety and Standards Authority of India (FSSAI) also regulates and defines novel foods as 'an article of food for which standards have not been specified but is safe for consumption; provided it does not contain any prohibited food and ingredients.'

PL technology is not widely used in the food industry. However, still, the Food and Drug Administration (FDA) in 1995 allowed the commercial-scale food packaging decontamination system based on PL. The FDA has approved PL treatment of foods in 1996 under code 21CFR179.41 (Code of Federal Regulations). US-FDA in 2020 also recommended that the fluence of PL should not exceed 12 J cm^{-2} for significant microbial or enzyme inactivation for food application. Like PL, UV is another non-thermal processing method much older than PL. According to CFR 21 part 179 (US-FDA 2001), the US-FDA considers UV radiation, ionizing radiation, radiofrequency radiation, PL, and CO_2 laser. It explained that if a radiation source is used to treat food, it is defined as a food additive. As per the food additive regulations, the product flow rate should follow a turbulent flow with a 'Reynolds number' above 2200. In 2013, based on the application and supporting documents, FSSAI also approved milk processing treated with UV.

PEF is another non-thermal process popular among researchers but not in industry. FDA does not have any specific regulations for PEF processing. Premarket approval is required before bringing any regulations. Thus, the FDA explains that a petition must be submitted for evaluation. Then a ruling will be made and issued by the Codex of Federal Regulations (CFR) after the assessment. Ozone processing is also gaining popularity as a non-thermal method. On 26th June 2001, the FDA approved ozone as an antimicrobial agent for direct contact with various foods and food products. Ozone processing also comes under 'generally recognized as safe' (GRAS). Ministry of Agriculture of the United States on December 2001 has also accepted the effect of ozone processing similar to an antimicrobial agent for use in direct contact with meats, poultry, fish, mollusks, and crustaceans. The Canadian

Food Inspection Agency (CFIA) has approved ozone processing for cleaning food contact surfaces. Few other countries (such as Japan, Australia, Germany, and Norway) agreed to its use in food, fishing, and aquaculture application.

Irradiation is one of the most popular non-thermal technologies used in the food industry, mainly for space foods. More than 40 countries have approved the application of irradiation on more than 50 food items. Initially, in the 1997 US Congress at Merrill, food irradiation was considered a food additive. The FDA approved irradiation on fruits and meat based on the concept of 'chemiclearance'. Joint (FAO, IAEA, and WHO) Expert Committee on Food Irradiation (JECFI) explained the 'Wholesomeness of Irradiated Foods', which helped draft its regulations. FAO stands for Food and Agriculture Organization, and IAEA stands for International Atomic Energy Agency. European Union (EU under Council Directive 1999/2/EEC), Codex General Standard for Irradiated Foods (No. 106-1983 adopted by the Codex Alimentarius Commission), and JECFI concluded that 'irradiation of food up to an overall dose of 10 kGy brings no nutritional or microbial issues and an overall dose over 10 kGy will begin to affect the nutrient content of food adversely'.

Contrary to this dosage limit, Codex Committee on Food Additives and Contaminants (CCFAC) agreed to remove the 10 kGy limits by bringing more practically applicable dose limitations. In July 2003, Codex Alimentarius Commission adopted this revision, stating, 'the maximum absorbed dose delivered to food should not exceed 10 kGy'. The energy sources for this irradiation process is regulated based on clause 2.1 of the Codex General Standard for Irradiated Foods. Ionizing radiations like gamma rays from ^{60}Co and ^{137}Cs, X-rays of 5 MeV or below, electrons of 10 MeV or below may be used for this irradiation process. In India, food irradiation is regulated by the Plant Quarantine Order (2004), the Food Safety and Standards Act (2006), and the Atomic Energy Rules (2012). Irradiated foods require special labeling with a *radura* logo (**Fig. 10.2**) and phrases such as *'treated with radiation to control spoilage'* or *'treated by irradiation for quarantine purposes'*. Manufacturers can identify the radiation source used, such as *'treated with gamma-radiation'*, *'treated with ionizing radiation'*, or *'treated with x-radiation'*.

HPP is the most popular and one of the industrially accepted non-thermal processing methods. In the United States, the Federal Food,

Figure 10.2 *Radura* logo for irradiation food products.

Drug, and Cosmetic (FD&C) Act, which regulates the processing, packaging, and preservation of foods along with the FDA, formulates the required regulations. Currently, high-pressure pasteurized products are supplied using refrigerated conditions. These should be processed under good manufacturing practices (GMP) conditions and follow product-specific regulations (e.g., juice hazard analysis critical control point (HACCP), Pasteurized Milk Ordinance (PMO), Sea Food HACCP, etc.). It also mentioned that the materials used in the equipment manufacturing and coming in direct contact with food items should be food-grade and approved by the relevant authority.

Cold plasma as non-thermal processing is popular among researchers and is yet to hit the market of non-thermal treated foods. Thus, this process is not regulated until now (2022), and it is recommended that the regulatory review process be approved for commercial cold plasma technology. OMFs are in the nascent stage of research. Thus, there is much work to be done before getting a nod from the regulatory bodies and bringing it into industrial practice.

10.6 Challenges and Research Needs

10.6.1 Equipment Cost

The initial investment for the equipment for a non-thermal process, such as HPP, PEF, cold plasma, PL, irradiation, and OMF, is one of the crucial obstacles. For any processed product, fixed cost, including the equipment, adds to the final product cost. For instance, one PEF equipment is producing 100 L/h of product. The plant operates 10 hours a day and 300 days a year. The initial equipment cost is USD 300,000, targeted to be recovered in 5 years. Therefore, the additional cost to the product due to equipment becomes USD 0.2/L (300,000/ [100 × 10 × 300 × 5]). The equipment is relatively cheap for UV, US, and ozonization. Therefore, designing a low-cost assembly for a specific non-thermal technique is obligatory.

10.6.2 Maintenance

The cost and resources related to the maintenance of non-thermal equipments are major concerns limiting their commercialization. For instance, a typical PEF electrode offers a life cycle of 100 h operation. The electrode is an expensive part of the PEF assembly. Similarly, in the case of HPP, the pressure intensifiers and high-pressure seals are prone to wear and tear, thus making the maintenance of the HPP process costlier. Therefore, research needs to address such issues so that non-thermal techniques can be easily commercialized.

10.6.3 Scale-up Trials

Most non-thermal techniques except HPP and PEF have been exploited in lab-scale equipment. The studies are required at least in pilot-scale levels so that the industry can adopt the technology. Continuous mode of operation is desirable for bulk processing, and this should be designed corresponding to a non-thermal technique.

10.6.4 Market and Consumer Acceptance

Consumers want a minimally processed and nutritionally superior food product that are fulfilled by the non-thermally treated product. However, the sensory analysis of the non-thermally treated product is the key. Besides, the product costs play a significant role here. The consumers are ready to buy the non-thermally treated product with a 20–25% extra cost as the thermally processed ones. This is the challenge for the manufacturer to come up with such pricing.

10.6.5 In-depth Understanding of Lethality

The mechanism of microbial or enzyme inactivation during non-thermal technology has not been concrete for some techniques. Some hypotheses have been accepted to describe some of those. For example, the mechanism of microbial inactivation during compression, pressure-hold, and decompression is not entirely established. However, it will help us design the time of HPP during compression and holding periods precisely.

10.6.6 Sustainability Aspect

The commercialization of the non-thermal technique depends on the sustainability aspects also. Most research on non-thermal techniques proved its potential in social aspects such as ensuring safety and nutritional quality. However, limited studies corroborated the environmental sustainability aspects. The energy efficiency also needs to be established for a non-thermal technique to be commercialized.

10.6.7 Type of Product and Matrix Properties

Several studies on non-thermal techniques are commodity-specific. In turn, the conditions are not applicable for other commodities. Besides, it is required to quantify the influence of matrix properties on the inactivation efficacy of a specific non-thermal technique. The product's suitability is also a determining factor for an industrial application. For instance, HPP is employed mainly on liquid or semi-solid products, whereas PEF is more suitable for fluid food. PL and cold plasma treatments are more effective for surface decontamination. Therefore, a product-specific optimized process condition for the different non-thermal techniques will be helpful. It will help the industry select a non-thermal technique for high-value low-bulk or low-value high-bulk commodities.

10.6.8 Shelf Life and Stability

The studies conducted so far showed that the shelf life of non-thermally treated products is not comparable with the equivalent thermally processed sample. One of the reasons might be the limited inactivation of microbial spores and enzymes in the non-thermal processed sample. In this sense, another hurdle may be added with the non-thermal technique. The hurdle may be in the form of any other non-thermal stress or mild thermal treatment in addition to the change in matrix properties, such as pH, sugar or salt concentration, etc.

10.7 Strategy for Process Optimization

Access to microbially safe and nutritious food for all people is one of the agendas to address social sustainability. That drives the food manufacturer to explore non-thermal technologies. Even if the food is microbially safe, browning, phase separation, and lipid hydrolysis in the food products by spoilage enzymes are not entertained. Thus, the retailers suffer the consequences and want a shelf-stable product. Most conventional thermal processes are always designed concerning microbial safety without considering enzyme inactivation. In turn, designing the non-thermal process satisfying the demands of all the stakeholders, viz. microbial safety (manufacturer and consumer want it) and nutrient retention (demand of consumers), and enzymatic stability (requirement from retailer) is the need of the hour.

This can be elaborated by taking an example of designing HPP for the pasteurization of pineapple juice. HPP is employed for pineapple juice to ensure microbial safety while retaining maximum phytochemicals. In addition, we need to ensure the inactivation of spoilage enzymes at that intensity. The experiments have been conducted at different combinations of pressure (P, MPa) and time (t, min) at a constant temperature of 40°C. The responses are target pathogen, enzyme, and bioactive components, such as *Escherichia coli*, polyphenol oxidase (PPO), and ascorbic acid (AA). A set of three contours can be plotted in a pressure-time landscape (**Fig. 10.3**).

The values for those contours are a 5-log reduction in *E. coli*, 99% inactivation of PPO and 10% loss in AA. For the socially sustainable high-pressure process design, our target will be safety ($\geq$ 5-log reduction of the target pathogen, *E. coli*) and stability (99% inactivation of the target enzyme, PPO) together with nutritional superiority (maximum 10% loss in target bioactive, AA). In this sense, at 500 MPa, 5 min treatment will ensure microbial safety, whereas 8 min treatment will show 99% PPO inactivation. From the contour for AA,

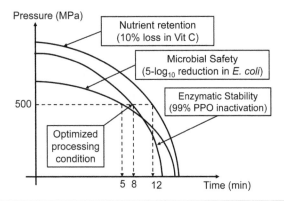

Figure 10.3 Strategy for designing a socially sustainable high-pressure process for pineapple juice at 40°C while satisfying consumers, manufacturers, and retailers' demand.

10% loss will be obtained for processing for 12 min at 500 MPa. For all these conditions, the temperature during HPP will be 40°C. Therefore, the optimized HPP condition for pineapple juice processing will be 500 MPa/40°C/8 min, which can satisfy all the demands of the stakeholders, viz. microbial safety (5-log reduction in *E. coli*) and nutrient retention (92% retention in vitamin C), and enzymatic stability (inactivation of polyphenol oxidase).

On a similar note, we need to design the non-thermal process intensity considering the three primary quality parameters of food processing technique, such as safety, stability, and quality. **Fig. 10.3** can be extrapolated for PEF, irradiation, and cold plasma (intensity vs. time), PL (fluence rate vs. time), US (acoustic power vs. time), ozonization (concentration of ozone vs. time), and OMF (magnetic field intensity vs. time). In turn, the process can be optimized concerning the food product's safety, stability, and quality. This will satisfy the goal of social sustainability in the food chain by providing safer food together with nutritional superiority. The generic nature of the new approach can be exploited for a wide range of food products.

10.8 Summary

- Non-thermal technology has the potential to overcome the disadvantages of thermal processing by achieving adequate safety, stability, with superior quality.
- The three main pillars of sustainability are economy, environment, and society.

- Social sustainability of non-thermal technology for food depends on factors, such as safety, nutrition, sensory appeal, nutrient bioavailability, shelf life, and need for preservative.
- Environmental sustainability depends on fossil fuel consumption, CO_2 emission, water consumption, waste utilization, and waste management of water and harmful chemicals.
- Economic sustainability is associated with the equipment cost, operation cost, maintenance cost, energy consumption, resource management, cost of research, and profit gains.
- The non-thermal equipment cost being higher is an issue, which is compensated by the market demand of the non-thermal technology.
- Proper awareness about the novel non-thermal technologies for food processing is necessary for gaining consumer acceptance.
- Food processed with HPP, PEF, and irradiation of food products, such as beverages, fruits, vegetables, meat, seafood, and ready-to-eat meals, are available in the market.
- Successful commercialization of the packaging material for the non-thermal processed food materials require a proper understanding of the product requirements, properties of the packaging material, authority regulations, and consumer response.
- Non-thermal processing needs deep study and optimization research to achieve the goal of food safety, stability, and quality, which will satisfy all the demands.

10.9 Multiple Choice Questions

1. Which of the following does not come under the environmental aspect of sustainability of non-thermal technology for food processing: (1) reduction of food waste, (2) food safety, (3) pollution control, (4) operational cost
   a. 1 & 2
   b. 1 & 3
   c. 2 & 4
   d. 3 & 4

2. What does the economic sustainability of HPP depend upon?
   a. Total cycle time
   b. Maintenance cost

   c. Energy consumption

   d. All of the above

3. Irradiation ($\geq$ 25 kGy) can't form volatiles when food is packed in _____.

   a. PE

   b. PP

   c. PET

   d. Glass

4. The regulatory body, named _____, formed the Joint Expert Committee on Food Irradiation (JECFI).

   a. FAO

   b. IAEA

   c. WHO

   d. All of the above

5. Which food authority includes all non-thermal treated food under the umbrella of novel food and processes category by stating that 'for novel foods and ingredients (CE258/97) established an evaluation, regulation, and a licensing system mandatory for all new foods and processes'.

   a. Food Safety and Standards Authority of India (FSSAI)

   b. European Union (EU)

   c. Food and Drug Administration of US (US-FDA)

   d. Canadian Food Inspection Agency (CFIA)

6. Which one of the following non-thermal techniques has not been operated in continuous mode?

   a. Pulsed electric field processing

   b. UV treatment

   c. Oscillating magnetic field processing

   d. High-pressure processing

7. Providing safety, stability, and food quality is a part of achieving _____.

   a. Economic sustainability

   b. Environmental sustainability

   c. Social sustainability

   d. All of the above

8. Which of the following is not a major challenge for the commercialization of non-thermal techniques?
   a. Equipment cost
   b. Comparable shelf life
   c. Maintenance
   d. Providing safety and quality

9. What was the main problem for the consumers regarding irradiated food that seemed to be preventing its commercial growth?
   a. Fear from the formation of carcinogenic components
   b. The problem of high product cost
   c. Uncertainty toward microbial safety
   d. Unsatisfactory nutritional and sensory qualities

10. Which packaging material is considered suitable for HPP processing of food?
    a. Paperboard
    b. Metal container
    c. Ethylene-vinyl alcohol (EVOH)
    d. Polyvinylidene chloride (PVDC)

10.10 Short Answer Type Questions

a. What are the necessary properties of the packaging material for the HPP of food?
b. Explain the labeling requirements for foods undergone irradiation.
c. With an example, describe the process optimization strategy for pasteurization of a fruit juice using PEF treatment.
d. Propose and discuss some strategies to make a non-thermal technique sustainable.
e. Highlight the challenges associated with the non-thermal processing of food products.

10.11 Descriptive Questions

a. Explain in detail, the three pillars of sustainability for non-thermal technology in food?
b. What should the regulatory body look into drafting regulation for a new product and process?

c. Discuss the packaging challenges associated with PL, irradiation, and PEF.

d. What should be the target for designing a pulsed light processing technique to pasteurize a fruit product?

e. Why is the shelf life of a non-thermally treated food product less than an equivalent thermally processed sample? How can it be improved?

References

Non-Thermal Processing Market for Food by Food Type, Technology, Region - 2022 | MarketsandMarkets. (n.d.). Retrieved April 23, 2021, from https://www.marketsandmarkets.com/Market-Reports/food-non-thermal-processing-market-158213636.html

Suggested Readings

Carreño, I. (2017). International Standards and Regulation on Food Irradiation. In I. C. F. R. Ferreira, A. L. Antonio, & S. C. Verde (Eds.), *Food Irradiation Technologies*. Royal Society of Chemistry, Milton Road, Cambridge, UK. pp. 5–27.

Chakka, A. K., Sriraksha, M. S., & Ravishankar, C. N. (2021). Sustainability of emerging green non-thermal technologies in the food industry with food safety perspective: A review. *LWT, 151*, 112–140.

Chauhan, O. P. (Ed.). (2019). *Non-thermal Processing of Foods*. CRC Press, Taylor and Francis Group, Boca Raton, Florida, USA.

Cullen, P. J., Tiwari, B. K., & Valdramidis, V. P. (Eds.). (2012). Status and Trends of Novel Thermal and Non-thermal Technologies for Fluid Foods. In *Novel Thermal and Non-thermal Technologies for Fluid Foods*. Academic Press, San Diego, USA. pp. 1–6.

Pereira, R. N., & Vicente, A. A. (2010). Environmental impact of novel thermal and non-thermal technologies in food processing. *Food Research International, 43*(7), 1936–1943.

Picart-Palmade, L., Cunault, C., Chevalier-Lucia, D., Belleville, M.-P., & Marchesseau, S. (2018). Potentialities and limits of some non-thermal technologies to improve sustainability of food processing. *Frontiers in Nutrition, 5*, 130.

Wani, F. A., Rashid, R., Jabeen, A., Brochier, B., Yadav, S., Aijaz, T., ... & Dar, B. N. (2021). Valorization of food wastes to produce natural pigments using non-thermal novel extraction methods: A review. *International Journal of Food Science & Technology, 56*(10), 4823–4833.

Watkins, D. (2012). Regulatory and Legislative Issues for Non-Thermal Technologies. An EU Perspective. In P. J., Cullen, B. K., Tiwari, & V. P., Valdramidis (Eds.) Novel Thermal *and* Non-Thermal Technologies *for* Fluid Foods. Academic Press, San Diego, USA. pp. 473–494.

Zhang, H. Q., Barbosa-Cánovas, G. V., Balasubramaniam, V. B., Dunne, C. P., Farkas, D. F., & Yuan, J. T. (Eds.). (2011). *Non-thermal Processing Technologies for Food*. Wiley-Blackwell, Chichester, West Sussex, UK.

Answers for MCQs (sec. 10.9)

1	2	3	4	5	6	7	8	9	10
c	d	d	d	b	d	c	d	a	c

Index

A

α-amylase, 172–173
α-helix, 68, 107
AA, *see* Ascorbic acid
Abiotic stress, 157
Absolute average error, 98
Absolute permittivity, 55
Absolute pressure, 118, 122
Absorbance, 70, 81
Absorbing species, 70, 82
Absorption coefficient, 71, 80, 160
AC, *see also* Alternating current
    AC magnetic field, 171
    AC voltage transformer, 175
Acid peroxides, 5, 7
AC-MF, *see* AC, AC magnetic field
Acoustic, 5, 7, 86–88, 90, 92,
        95, 207
    acoustic cavitation, 5, 7
    acoustic energy, 87
    acoustic field, 86
    acoustic frequencies, 88, 95
    acoustic intensity, 92
    acoustic power, 90, 207

Acoustic power density, 87, 97
Activated carbon, 129
Activated water, 133, 140, 192
Activation energy, 27, 35, 39, 99
Activation volume, 27, 33–35,
        37, 39
Active site, 25, 68, 89, 102, 148
Active species, 112–113, 121
Active zone, 114, 121
    active zone species, 114
Adiabatic compression, 23
Adiabatic cooling, 22–23
Adiabatic heat, 22–23, 190
Advanced thermal processing, 3
Aerobic, 29, 40, 77, 93, 122, 151
    aerobic bacteria, 151
    aerobic mesophiles, 29, 40, 77
    aerobic mesophilic count, 29,
        40, 77
    aerobic psychrotrophic
        bacteria, 122
Aflatoxins, 108, 114, 120, 123,
        146, 161
Agrochemical residues, 114
Air bubbles, 190–191

Air molecules, 119
Air-core solenoids, 174, 180
Alkaline earth metals, 68
Alkaline phosphatase, 47
Allergens, 10, 108, 114, 120, 192, 194–195
Alloys, 90
ALP, *see* Alkaline phosphatase
Alpha particles, 143
Alternaria solani, 182, 187
Alternating current, 11, 43, 48, 90, 101, 109, 121, 167, 171, 175, 177, 180–181
Alternating magnetic fields, 167, 171
AM, *see* Aerobic, aerobic mesophiles
Ambient, 30, 149, 156, 190
    ambient conditions, 149, 156
    ambient pressure, 190
    ambient temperature, 30, 156
AMC, *see* Aerobic, aerobic mesophiles; Aerobic, aerobic mesophilic count
AMF, *see* Alternating magnetic fields
Amino group, 5, 7, 26, 30, 47, 68, 89, 102, 107, 116, 147–148, 162, 173
Amla, *see* Indian gooseberry
Ampere's law, 181–182
Anaerobic bacteria, 93, 151
Anisotropy, 172–173
Anthocyanin, 89, 196
Anti-carcinogenic, 179
Antimicrobial, 127, 131, 133–134, 140, 179, 201
Antimicrobial action, 131, 140
Antioxidant capacity, 7, 29, 47, 53, 68, 178, 191
AOC, *see* Antioxidant capacity
APD, *see* Acoustic power density
Apple, 9, 70, 94, 104, 116
    apple cider, 9
Aquaculture, 202

Aqueous ozone, 130, 132, 134
    aqueous ozone treatment, 130, 132, 134
Ar, *see* Argon
Arc, 68, 77–78, 105, 109, 115, 121
Arginine, 162
Argon, 66, 68, 81, 105, 115–116
Aroma, 128, 133
Aromatic, 173
Arrhenius, 27, 35, 100
    Arrhenius-Eyring, 35, 39
Artificial ozone, *see* Ozone gas
Ascorbate oxidase, 195
Ascorbic acid, 29, 48, 53, 57–59, 62, 135, 206
Aseptic, 3, 19, 29, 169
    aseptic design, 29
    aseptic environment, 19
    aseptic processing, 3, 169
Atmospheric pressure, 17, 24, 29, 105, 109, 115, 136
Atomic energy, 11, 196, 202
Atomic mass, 145, 163
Atomic number, 145, 151
Atomic oxygen, 107
Attenuate, 87, 92, 160
Avure Technologies, 198

B

β-carotene, 148
β-sheet, 47, 68, 107
Bacterial spore, 127, 192
Bactericidal, 11, 112
Barium titanate, 90
Baro-resistant, 37
Batch operation, 18, 20, 26, 28, 31, 39, 49, 54, 68–69, 76, 190
Bats, 85
Beef, 108, 158–159, 169
    beef carcass, 158–159
    beef fillets, 159
Beer-Lambert law, 70, 80–82

Berries, 150

Beta rays, 145–146, 151

Beverage, 8, 10, 30, 47, 57–58, 62, 75, 132, 208

Biaxial oriented polypropylene, 200

Binding site, 171–173

Bioactive compounds, 7, 9, 26, 40, 47, 81, 96, 115, 155, 194, 206

Bioavailability, 48, 190, 194, 208

Biomass, 180

Biomolecules, 145, 172

Bioreactors, 88

Biotechnology, 9

Bipolar, 51–52, 55

  bipolar pulses, 51

Black box system, 173, 185

Blueberry, 131–132, 141

Boiling point, 127

Boltzmann constant, 118

BOPP, *see* Biaxial oriented polypropylene

Bosch, 198

Bottom seals, 20, 23

Bread, 190

Brix, 53, 74

Bromelain, 39

Browning, 30, 53, 74, 206

Bubble diffuser chamber, 129, 139

Bubble diffusion system, 135

Budding, 172

Buffer solution, 83, 97

Bühler, 198

Bulk processing, 29, 204

Burning, 195

Bursts, 4, 43

By-products, 53, 195, 197

C

Ca, *see* Calcium

Caesium, 151–152

CAGR, *see* Compound annual growth rate

Calcium, 170–172

Calcium-protein bond, 172

Calmodulin, 171–172

Canadian Food Inspection Agency, 201–202, 209

Canning, 2

Cantaloupe, 174

Capacitance, 49, 54–55, 61, 178, 190

Capacitor, 43–44, 48–49, 69, 174–175, 178

  capacitor bank, 48, 175, 178

Capillary hemorrhage, 133

Capital cost, 30, 190, 196

Capsanthin, 156

Carbohydrate, 47, 107–108, 120, 173, 178

Carbon, 54, 129, 139, 174, 196

  carbon dioxide, 139

  carbon electrodes, 54

  carbon emissions, 196

Carboxyl group, 47

Carbs, *see* Carbohydrate

Carcinogens, 148

  carcinogenic, 114, 198, 210

Carotenoids, 62, 151

Carrot, 74, 83

Catalytic decomposition, 129

Catalytic power, 25

Catechin, 178

Cathode, 151–152

  cathode rays, 11, 151

Cavitation, 5, 7, 13, 85–90, 92–93, 95–97, 101–103

  cavitation bubble, 85–89, 93, 95–97, 101–103

  cavitation threshold, 85, 95

CCFAC, *see* Codex, Codex Committee on Food Additives and Contaminants

Cell hybridization, 9

Cell lysis, 7, 46, 192
Cellular membrane, 107
Cell organelles, 46, 134, 147, 170, 177
Cell wall, 5, 7, 88, 93, 96, 113, 127
Cellular protein, 170, 178
Cellulose, 162
Cereal, 149
Cereal-based formulations, 8
Cesium, 11
CFIA, *see* Canadian Food Inspection Agency
CFR, *see* Codex, Codex of Federal Regulations
Charged species, 116
    charged molecule, 143, 147
    charged particle, 10, 105, 107, 115, 143, 145, 170–171, 184–186
Chemical bonds, 65, 87, 145
Chemical energy, 106, 115
Chemical preservatives, 52, 195
Chemical reaction, 25, 107–108, 119, 127, 138, 149
    chemical activity, 173
Chemiclearance, 202
CHIC FreshTech, 198
Chick-Watson model, 131, 140–141
Chili, 155–156
Chitin, 113
Chocolate, 92
Circular economy, 194
Citric acid cycle, 148
Climacteric fruits, 157
Clonogenic death, 66, 82
Co, *see* Cobalt
CO_2 emission, 195, 208
Coaxial chamber, 51
Coaxial cylindrical electrodes, 60
Coaxial electrode design, 50
Cobalamin, 148
Cobalt, 11, 14, 90, 151–152, 163
Coconut water, 70, 117–118

Codex, 201–202
    codex alimentarius commission, 202
    Codex Committee on Food Additives and Contaminants, 202
    Codex of Federal Regulations, 201
Cohesiveness, 87, 191
Collinear chamber, 51
Color pigments, 75
Come-up time, 18, 23, 28
Commercial practices, 9
Commercial sterility, 149
Compound annual growth rate, 8
Compression rate, 23, 37
Compton effect, 144, 163
Compton scattering, *see* Compton effect
Concentric electrodes, 111
Condensation, 30, 129, 139
Condenser, 49
Conduction, 4
Conductivity, 44, 49, 57, 59, 61–62, 177, 180, 187, 190
Conductor, 167, 180, 184
Consumer acceptance, 200, 208
Consumer demand, 194, 197
Consumer education, 157
Contact surface, *see* Food contact surface
Contactor, 128, 139
Contaminate, 52, 192, 200, 202
Continuous parallel plate system, 51
Continuous processing, 44, 50, 54, 190, 204, 209
    continuous HPP, 18, 39
    continuous PL, 69
    continuous ozone treatment, 132
    continuous UV, 9, 79
    continuous-type PEF system, 43
Controlled atmospheric conditions, 130, 178

Convection, 4, 187
Conventional pasteurization, 195
  conventional thermal
      pasteurization, 9, 53, 197,
      200, 206
Cooking, 108, 120, 195
Cooling, 13, 22, 48, 95, 105, 130,
      193
  cooling system, 48, 95
Copper, 101, 178
Coriander, 155–156
Corona discharge system, 125–126,
      129, 134, 138–140
  corona discharge ozone generator,
      138
  corona discharge plasma, 109,
      120
  corona discharge type generators,
      129
Corrosion, 95, 129, 133,
      139, 191
Coupling energy, 170, 172, 180
Covalent bonds, 5, 7, 25–26, 31,
      172, 190
  covalent cross-linkages, 68
Critical electric field, 46
Cross-contamination, 28, 200
Crustaceans, 201
Crystals, 174, 193
Cumulative lethality, 31
Curcumin, 156
Current density, 115, 169
Cyclotron, 169–170, 172
  cyclotron frequency, 172
    cyclotron angular frequency,
      170
    cyclotron resonance, 169–170
      cyclotron resonance frequency,
      170
Cysteine, 68, 147–148, 162
Cytoplasm, 46
  cytoplasmic, 5–7, 67
Cytosine, 66

D

Dairy, 8, 10, 30, 132, 156
  dairy products, 8, 30
Dalton's law, 38
DBD, *see* Dielectric barrier
    discharge
  DBD-based cold plasma, 115,
    117, 122
DC, *see* Direct current
  DC-MF, 171
Deactivation, 11
Death kinetics, 179
Death phase, 50, 177
Decarboxylation, 148
Decaying plasma, 111
Decimal reduction dose, 146, 155
Decimal reduction time, 93–94, 97,
    99, 122, 146
Decomposition derivatives, 127, 134
Decompression time, *see*
    Depressurization
Decontamination system, 201
Defoaming, 9, 191
Degassing, 10, 87, 191
Degradation rate, 58, 62, 135–137
  degradation rate constant, 58, 62,
    135
  dehydrated milk, 146
Dehydration, 13, 53, 191
Dehydrogenase, 148
  dehydrogenase activity, 148
Deinococcus, 193
Deionized water, 187
Denaturation, 5, 7, 25, 116, 148
Deodorization, 10
Depolymerization, 68, 81, 149,
    157–158, 193
Depressurization, 19, 23–24, 28, 31
Deprotonation, 150
Depth dose distribution, 154, 157,
    159
Detection limit, 29

Diamagnetic, 173
Diatomic oxygen, 10, 126
Dielectric constants, 44
Dielectric barrier discharge, 5–6, 10, 109–110, 113, 115–117, 121–122, 192
    dielectric barrier discharge plasma, 110, 121
Dielectric breakdown, 119, 190
Dielectric constant, 44, 49, 55–56, 59–60
Dielectric discharge, *see* Dielectric barrier discharge
Dielectric heating, 3
Dielectric rupture theory, 44–45, 54
Differentiation, 160
Diffuse plasma, 110
Diffusion, 48, 85, 101, 135
Digestibility, 148, 190, 194
Diode plasma discharge, 112
Dipolar molecules, 49
Dipole, 46, 172
    dipole nature, 46
Direct current, 48, 167, 171, 174, 180
Discharge volume/area, 110, 120
Discoloration, 192
Disinfect, 75, 191
    disinfectant, 116, 133, 192
Disintegrate, 81, 88, 191
Dissociative recombination, 107–108, 121
Disulfide bond, 25, 47
DL, *see* Detection limit
DNA denaturation, 116, 180
    DNA photolyase, 67
DNA repair system, 151, 193
Dose, 67, 74, 76, 79, 83, 145–146, 150, 154–160, 162, 164, 200, 202
    dose of irradiation, *see* Gray
    dose rate, 156

dose uniformity ratio, 154–155, 159–160, 163
Dosimeter system, 154
    dosimeters, 154
Dry cereal powders, 149
Dry fruits, 190
Drying, 191, 195
DUR, see Dose, dose of irradiation; Dose, dose uniformity ratio
D-values, *see* Decimal reduction time
Dwelling time, *see* Holding time
Dynamic phases, 28

E

Earth's MF, 172, 177
Economic sustainability, 189, 196, 208–209
Edible oil, 10
Effective outer surface area, 77
Elderberry, 196
Electric gun, 152
Electrical excitation, 68
Electrical insulators, 49
    electrical resistance, 176
    electrical resistivity, 59, 193
Electricity, 175, 198
    electric current frequency, 122
    electric field lines, 45
    electric field strength, 43, 50, 57, 60, 62, 169
    electric oscillation, 90
    electric power, 101, 128
    electrical breakdown, 5, 7, 105
    electrical conductivity, 59, 177, 180
    electrical discharge, 10, 115
    electrical energy, 43, 60, 69, 89, 96, 196
Electrochemical reaction, 54, 190
Electro-compression, 45–46, 59

Electro-conformational, 47–48, 54, 59
Electrode corrosion, 129, 139
Electrolysis, 9, 126–127
Electrolytic methods, 125, 133
    electrolytic decomposition, 126, 140
Electromagnet, 11, 174–175, 177, 181
Electromagnetic energy, 196
Electromagnetic transducers, 90, 102
    electromagnetic wave spectrum, 143–144
Electromagnetic waves, 110, 143, 145, 167, 185
    Electromagnetism, 182
Electron accelerator, 152
Electron beam emissions, *see* Cathode, cathode rays
    electron beam irradiation, 153, 161–162
    electron beams x-rays, 143
Electronvolt, 65, 105, 145, 172
Electroporation theory, 44–45
    electroporation effect, 119
Electrostatic, 25
Electrostrictive, 90
Electrostrictive property, 90
EM wave, *see* Electromagnetic waves
Empirical model, 35
Emulsification, 9
Energy density, 55, 113
    energy density distribution, 55
Enumeration, 156
Environmental sustainability, 189, 205, 208–209
Enzymatic inactivation, 121
Enzymatic stability, 12, 83, 206–207
Enzymatic-browning, 190
Enzyme binding site, 171, 173
Enzyme conformation, 25

Enzyme proteins, 47, 116, 173
    enzyme's active site, 68, 89
Enzymes' residual activity, 117
Ervinia carotovora, 182, 187
Etching, 108, 120
Ethyl alcohol, 132
Ethylene-vinyl acetate, 200
Ethylene-vinyl alcohol, 24, 199, 210
Ethylene-vinyl alcohol copolymer (EVOH), 24
EU, *see* European Union
European Union, 201–202, 209, 211
eV, *see* Electronvolt
EVA, *see* Ethylene-vinyl acetate
EVOH, *see* Ethylene vinyl alcohol
Excitation, 68, 131, 171
    excited atom, 107
    excited electrons, 65
Excited state, 10, 105
Explosion energy, 78
Exponential decay, 51, 55, 60
    exponential decay waveform, 55, 60
Exponential growth phase, 113, 130
Extinction coefficient, 70–71, 78–79, 82
Extraction, 9–10, 95, 191, 196, 211
    extraction techniques, 95
Extrinsic factors, 148, 162, 164
Eyring's equation, 27, 33, 35

F

FAO, *see* Food and Agriculture Organization
Fatty acids, 5, 7, 47, 127, 148
FDA, *see* Food and Drug Administration
FD&C, *see* Food Drug and Cosmetic Act
Fermentation, 180
Ferromagnetic materials, 90, 101–102

Fiber, 173

Filtration, 10, 191

First-order kinetic model, 28, 58, 77, 104, 117, 131, 187

first-order decay, 155

first-order inactivation model, 33

Fish, 10, 26, 122, 132, 149, 201

fish fillets, 122

Flash lamp, 66, 69, 77, 81

Fluence, 69, 71–74, 76–77, 79–83, 201, 207

fluence rate, 71–73, 76–77, 80–83, 207

Fluid food, 18, 84, 191, 205, 211

Flux densities, 175, 181

Flux of magnetic field, 171

Food allergens, 10, 108, 114, 120, 192, 195

Food and Agriculture Organization, 11, 19, 41, 202, 209

Food and Drug Administration, 201–203, 209

Food contact surface, 4, 9–10, 133, 191, 198, 202

Food contamination, 95, 115, 191

Food Drug and Cosmetic Act, 202–203

Food industry, 123, 142, 196, 201–202, 211

food chain, 198, 207

Food matrix, 26, 48, 72, 92, 149–150, 155

Food preservation methods, 5, 169

food packaging, 200–201

food storage, 130

Food quality, 9, 95, 158, 169, 174, 190–191, 193, 209

flavor components, 17

food flavor, 133

Food Safety and Standards Authority of India, 201, 209

Food spoilage, 119, 128, 150

Food waste, 196, 208

Foodborne pathogens, 150, 193, 195

Forces of interaction, 89, 102

Forged monolithic wall, 20

Fossil fuel, 195, 208

Fourth state of matter, 10, 105

Freezing point, 22, 187

Frozen foods, 132, 150, 174

frozen beef, 158

frozen seafood, 158–159

Fruit product, 1, 211

fruit juice, 26, 40, 49, 52, 74, 83, 210

fruit juice-milk beverages, 47

fruit purees, 40

Fruit ripening, 148

FSSAI, *see* Food Safety and Standards Authority of India

Fumigation, 132, 149

Functional groups, 147

Fungal spores, 27, 71, 76, 80, 93, 113, 138, 190–191

Fungal toxins, 146, 193

Fuzzy analysis, 40

G

γ, *see* γ-rays

γ-irradiation, 151, 153, 155–156, 193, 196

γ-rays, 193

GAD, *see* Gliding arc, gliding arc discharge

GAE, *see* Gallic acid, gallic acid equivalent

GAEAC, *see* Gallic acid, gallic acid equivalent antioxidant capacity

Gallic acid, 29, 53, 74

gallic acid equivalent, 29, 53, 74

gallic acid equivalent antioxidant capacity, 29

Gamma emission, 152
  gamma irradiation, 143,
    145–146, 148, 153–154,
    161, 164
  gamma rays, 11, 14, 144–146,
    157, 163, 185, 202
Gas constant, *see* Universal gas
    constant
Gating potential, 46, 54
Gel-like structure, 46
Gelling power, 157
General oscillator's inertia, 92
Generally recognized as safe, 201
Genetic information, *see* Genetic
    material
Genetic material, 5, 7, 81, 146–147,
    157, 193
Genotoxicity, 149
Germicidal radiation, 70, 76
  germicidal property, 66
Germination, 114
Glass containers, 180
  glass bottles, 24, 200
Gliding arc, 109, 121
  gliding arc discharge, 109, 113,
    116, 121
    gliding arc discharge plasma,
    109, 121
Globular structure, 173
Glycolipids, 127, 134
Glycoproteins, 127
GMP, *see* Good manufacturing
    practices
Good manufacturing practices, 203
Gradient MF, 174
Gram-negative cells, 93
  gram-negative bacteria, 26, 37,
    49, 71, 80, 88, 93, 101, 131,
    151, 190
  gram-negative pathogen, 141
Gram-positive cells, 93
  gram-positive bacteria, 37, 49, 71,
    80, 88, 93, 101, 113, 151

Grapefruit juice, 62
GRAS, *see* Generally recognized as
    safe
Gray, 145, 150, 154, 156
Ground state, 10, 105
Ground state, 10, 105, 164
Grounded chicken, 164
Groundnut, 114, 123
Growth medium, 94, 114, 131
Growth phase, 50, 113, 130, 177
Gy, *see* Gray
Gyrofrequency, *see* Cyclotron,
    cyclotron frequency,
    cyclotron angular
    frequency

H

5-hydroxymethyl-2-furfural, 53
HACCP, *see* Hazard analysis and
    critical control point
Half-life, 136–137, 152, 182–183,
    187
Half-value layer, 160–161
Ham, 26
Hazard analysis and critical control
    point, 203
HDPE, *see* High-density
    polyethylene
He, *see* Helium
Headspace, 199
Heat energy, 1, 4, 6
Heat exchanger, 44, 74
Heat of compression, 6, 22, 37, 60,
    190
Heat resistance, 200
Heat transfer rate, 3
Heavy metal, 20
Helium, 105, 115, 143, 175
Henry's law, 141
High impact polystyrene, 200
High-density polyethylene, 200
High-pressure pumps, 19

High-temperature processing, 24
  high-temperature short time
    technique, 3
High hydrostatic pressure (HHP),
  208
Hiperbaric España, 198
HIPS, *see* High impact polystyrene
Histidine, 147
HMF, *see*
        5-hydroxymethyl-2-furfural
Holding time, 18–20, 23, 28, 31,
    40, 205
Homeostasis, 6
Homogenization, 13
Horseradish, 173
Hot water bath, 52
HTST, *see* High-temperature
        processing, high-
        temperature short time
        technique
Human colonic cells, 149
Hurdle techniques, 75–76
HVL, *see* Half-value layer
Hybrid magnet, 175, 181
Hydration capacity, 26
Hydrogen bonds, 7, 25, 89,
    102
Hydrogenation, 10, 108
  hydrogenated oil, 78
Hydrolysis, 206
Hydroperoxide radical, 127, 134
Hydroperoxyl radical, 147
Hydrophilic, 46, 59
Hydrophobic, 26–27, 46
Hydrostatic pressure, 4, 17, 30, 39,
    41
Hydroxyl species, 89
  hydroxyl radical, 87, 96, 107,
      113, 127, 134, 147,
      149
Hyperbaric pressure processing,
    8, 17
Hypotonic solution, 89

I

IAEA, *see* International Atomic
      Energy Agency
ICR, *see* Ion cyclotron resonance
ideal gas, 118, 122
  ideal gas law, 118
Ignition voltage, 109, 121
Immiscible liquids, 78
Immunecompromised, 149
Incomplete, 88, 102, 105,
      119–120
  incomplete ionization, 105, 120
    incomplete ionization pathway,
      120
Indian gooseberry, 52–53
Indicator microorganism, 180
Indirect pressurization, 21, 31, 39
Induced mutations, 5
Inductive reactance, 181
Inert gas, 66, 68
Infant foods, 30
Infinite slab, 187
Infrared, *see* IR
Initial enzyme activity, 99
Initial microbial count, 28, 50, 56,
      72, 94, 113, 130
Initial microbial load, 27, 50, 72, 93,
      155
Inlet feed gas, 128,
Inner electrode, 111
Inoculation, 52, 56, 62, 94, 97, 114,
      116, 131–132, 141, 164
Inorganic molecule, 10
In-pack treatment, 28, 179, 190,
      193, 200
Intermolecular disulfide
      bond, 25
Internal energy, 171, 196
International Atomic Energy
      Agency, 11, 202, 209
Intracellular, 25, 44
Intrinsic, 27, 71, 148, 150, 164

Ion cofactors, 171, 173
Ion cyclotron resonance, 170, 173, 180
Ion parametric resonance, 171, 173, 180
Ion translocation, 180
Ion-bound protein, 172
Ion-dependent reactions, 173
Ion-enzyme complex, 173
Ionic bonds, 25
Ionic strength, 141
Ionization, 6, 105–107, 110, 115–116, 119–120
    ionized plasma, 106, 115
    ionizing irradiation, 149, 151, 157, 162–64
    ionizing radiation, 143, 145–146, 151, 155, 157, 163, 165, 189, 193, 201
    ion-protein bonds, 170
IPA, *see* Isopropyl alcohol
IPR, *see* Ion parametric resonance
IR, 3, 7, 67, 73, 85, 145
Iron-core inductor, 175
Irradiation dose, 150, 159, 164
Isobaric-isothermal inactivation, 39
Isopropyl alcohol, 22
Isostatic, 4–5, 17, 21, 30, 38
Isotonic solution, 102
Isotope, 145, 163, 173–174
Isotropy, 173, 180

J

JECFI, *see* Joint (FAO/IAEA/WHO) Expert Committee on Food Irradiation
Jet-type cold plasma equipment, 116
Joint (FAO/IAEA/WHO) Expert Committee on Food Irradiation, 202, 209
Joule heating, 43, 46–49, 54, 174, 180–181, 186, 193

K

Kinetic energy, 106, 144–145
Kinetic equation, 50, 72, 77, 99
Kinetic model, 27–28, 50, 58, 62, 104, 117, 131, 135, 141, 164
Kr, *see* Krypton
Krypton, 66, 81
K-type, 22

L

LAB, *see* Lactic acid bacteria
Labeling, 156, 202, 210
Lactate dehydrogenases, 148
Lactic acid bacteria, 122
Lag phase, 50, 177
Laminated plastic films, 199
Lamp, 65–66, 68–69, 73, 76–79, 81–82
Latent heat of fusion, 187
LDPE, *see* Low-density polyethylene
Le-Chatelier's principle, 17, 38
Lead (Pb), 161
Legumes, 108
Leucine, 162
Licensing, 201, 209
Life cycle, 134, 189, 204
Lifespan, 77–78, 197
Light energy, 71–73, 76, 79, 81, 83, 196
    light energy flux, 72
Light intensity, 77, 80–81
Light spectrum, 65–66
Light transmittance, 191
Linearized Arrhenius equation, 100
Lipase, 147, 164
Lipid, 17, 46–47, 59, 108–109, 115, 148–149, 158, 173, 192–194, 206
    lipid bilayer, 46–47, 59

lipid hydrolysis, 206
lipid oxidation, 108, 149, 192
Lipopolysaccharide, 113
Local minima, 159–160
Localized heating, 5, 7
Logarithmic absorption coefficient, 71
Logarithmic destruction ratio, 83
Logarithmic growth phase, 50
Logarithmic survival fraction, 141
Log cycles, 33
Log-linear, 40, 83
Long-lived, 111, 113
Longitudinal wave, 85–86, 101
Lorentz force, 170
Low-density polyethylene, 200
Lysis, 7, 46, 192

M

Machinery, 76, 115
Macronutrients, 148
Magnet, 11, 90, 101–102, 167, 174–175, 180–181, 193
Magnet-based OMF system, 186
Magnetic coil, 6, 175, 181
Magnetic field intensity, 181–182, 207
Magnetic flux, 167, 169, 175, 180–182, 185–187
  magnetic flux density, 167, 169, 175, 180–182, 185–187
Magnetic freezing, 174, 186
Magnetic particles, 173
Magnetic permeability, 184
Magnetic properties, 90
Magnetic susceptibility, 173–174, 180
Magnetically active component, 167, 170, 172
Magnetism, 4, 11, 167
Magnetization, 167, 173

Magnetostrictive, 89–90, 96, 102
  magnetostrictive transducers, 90, 102
Maillard reaction, 30
Malate dehydrogenase activity, 148
Mango pulp, 27
Manothermosonication, 83
Mechanica, 6, 86, 89–90, 96, 171–172, 196
Melanoidin, 30
Membrane conductivity, 44
Membrane integrity, 46, 61
Membrane permeability, 46
Membrane rupture, 9, 44–45, 59
Membrane rupture theory, 9
Mesophiles, 29, 77
Mesosomes, 67
Metabolic activities, 47, 171
Metabolic disorders, 157
Metabolic proteins, 5, 7
Metabolic reactions, 148, 171
Metabolism, 25, 177
Metabolites, 114
Methionine, 68, 162
Methylesterase, 27, 47, 74, 89
MF-based microbial inactivation, 169, 186
Microbial analysis, 122
  microbial decontamination, 10, 114
  microbial inhibition, 11
  microbial invasion, 13
  microbial safety, 3–4, 12, 14, 30, 83, 195, 206–207, 210
  microbial spoilage, 2
Microflora, 29, 150, 159
Micronutrients, 148, 194
Microwave, 3, 112, 115, 121, 145, 169, 185, 192
Moisture content, 130, 149, 190
Molar absorptivity, 70
Mold, 29, 37, 95, 114, 156, 190
Momentary joule heating, 47–48, 54

Monochromatic, 68, 76
Monolithic, 20
Monopolar, 51–52
Mono-propylene glycol, 22, 38
Morphology, 25, 192
Mortality, 1
MPG, *see* Mono-propylene glycol
Mutagenic, 133, 149
Mutations, 5, 7
Mycotoxins, 114, 128, 131–132, 134,
    192, 195

N

Nascent, 203
Ne, *see* Neon
Near-infrared, 66, 76
Nefarian absorption coefficient, 71,
    80
Neon, 81
Neutrons, 145
Niacin, 148
Nitrates, 148
Nitric acid, 108, 112
Nitrites, 148
Nitrosamines, 148–149
Noble gas, 81, 115, 191–192
Nonenzymatic browning index, 53,
    74
Nonequilibrium, 105, 115, 120–121
Nonionizing radiations, 145
Nonprotein, 171
Nonradioactive, 152
Non-recombination pathway, 121
Nontoxic, 10, 125, 133, 140, 191
Nordion, 198
Normalized, 178
Novel food, 123, 201, 209
Novel processing, 123, 200, 208,
    211
Nuclear reactor, 105
Nucleic acid, 26, 151
Nylon PP pouches, 199

O

Ochratoxins, 146
Odor-causing volatiles, 128, 134, 192
Off-flavor, 148–149
Ohm's law, 181
Ohmic heating, 3, 43, 181
Online-cooling mechanism, 75
Oocyst, 134–135
OPP, *see* Oriented polypropylene
Optical sensor, 69
Optically opaque, 70, 191
Optimization, 40, 53, 197, 208, 210
Orbital electron, 144–145, 162
Organic, 53, 130, 148, 174, 193
Organoleptic, 47, 53, 68, 197
Oriented polypropylene, 199–200
Oscillation, 90, 172
Osmolytes, 26
Osmotic imbalance, 46
Overheating, 4, 69
Oxygen gas, 125, 133, 174, 197
    oxidative, 5, 7, 115, 128, 149, 179,
        195
    oxidizing, 10, 127–128, 133–134,
        147, 195
    oxygen atoms, 125–126
Ozone gas, 6–7, 10, 125, 129–130,
        132, 136
    ozonated, 6, 141
    ozonation, 5, 10, 127
    ozone concentration, 128–129,
        131–136, 138, 141
    ozone decay, 130, 139, 192
    ozone generation, 125, 129, 134
    ozone generator, 128, 132, 138
    ozone processing, 125, 128–129,
        131–132, 140, 192–193,
        201–202
    ozone treatment, 4, 12, 128–134,
        139, 142, 192, 195
    ozone-activated water, 140–141
    ozonization, 6, 13, 204, 207

P

Papain, 148
Parallel plate, 50–51, 55, 60
    parallel plate electrodes, 60
    parallel plate system, 51
    parallel type electrode, 55
Paramagnetic, 172–174
Parametric, 171
    parametric excitation, 171
Parasites, 134, 145, 190
Pascalization, 8, 13–14, 17
Pasteurized milk ordinance, 203
Path length, 70, 82, 161
Pb, *see* Lead (Pb)
PE, *see* Polyethylene
Pectin methylesterase, 27, 32–34, 47,
    74–75, 89, 98–100
Penetrability, 10, 125
Peptide, 68, 148, 172
Peptide bonds, 68
Peptidoglycan, 6, 93, 113, 128
Peptone water, 92
Permanent magnet, 90, 101–102,
    167, 180
Permeability, 5, 9, 13, 25, 44, 46, 61,
    184, 200
Permittivity, 55
Peroxidase, 27, 47, 117–118, 173
Peroxide, 5, 7, 107, 112, 147, 157,
    193
Perturbation, 46, 90
Pest management, 149, 162
Pesticide, 10, 108, 120, 127–128,
    131–132, 134, 161,
    192, 195
PET, *see* Polyethylene terephthalate
PG, *see* Polygalacturonase
Phenolic content, 53, 68, 74, 104,
    178, 191
Phenylalanine, 68
Photoelectric effect, 144, 162
Photolyase, 67

Photon, 65–66, 107, 120, 143–145,
    157, 162–163, 196
Photo-oxidation, 68, 81
Photophysical, 67, 80, 82
Photoreactivation, 67, 76
Photo-reactivation, 81
Photo-reduction, 81
Photothermal, 7, 67, 71, 76,
    80, 138
Physicochemical, 74, 83, 173–174
    photochemical effect, 7, 65–68,
    71, 76, 81, 119
Phytochemical, 7, 14, 29, 31, 39,
    194, 206
Piezoelectric, 89–90, 96, 101
Pigments, 26, 75, 191, 211
Pin to plate type plasma, 112
Plank's constant, 65, 143
Plank's equation, 187
Plank-Einstein's equation, 65, 82
Plant quarantine order, 202
Plasma decay, 106, 115
Plasma decontamination, 123
Plasma device, 106, 109, 112, 115,
    120–122
Plasma generation, 105,
    120, 122
Plasma jet, 109, 111, 115, 192
Plasma plume, 111
Plasma processing, 105, 114–115,
    118, 121, 192
Plasma species, 108, 111, 113
Plasma technology, 10, 123, 203
Plasma treatment, 5, 107, 112, 114,
    121–122, 196, 205
Plasma-activated water, 111–112,
    121, 192
Plastic, 24, 122, 180, 199–200
Plated agar, 94, 114, 131
PLC, *see* Programmable logic
    controller
PME, *see* Pectin methylesterase
PMF, *see* Pulsed magnetic fields

PMO, *see* Pasteurized milk ordinance
POD, *see* Peroxidase
Polarity, 168, 175, 180
Polarization, 44–45, 49, 90, 101
Polychromatic light, 9, 66, 68, 76
Polyester, 199
Polyethylene, 199, 200, 209
Polyethylene terephthalate, 122, 199–200, 209
Polygalacturonase, 47
Polyketide, 114
Polymerization, 148
Polyols, 26
Polypeptide, 89, 102
Polyphenols, 47, 68, 128, 133, 179
    polyphenol oxidase, 27, 29, 47, 104, 117–118, 206–207
Polypropylene, 199–200
Polysaccharides, 26
Polystyrene, 200
Polyunsaturated fatty acids, 5, 7, 127
Polyvinyl alcohol, 24, 199
Polyvinylidene chloride, 200, 210
Positron, 145
Post-processing, 1, 12, 28, 31, 200
Poynting vector, 167
PPO, *see* Polyphenols, polyphenol oxidase
Preconditioning, 44, 49
Preheating, 98–99, 197
Pressure vessel, 18–21, 30, 38
Pressure-transmitting medium, 18–19, 21–23, 38
Process design, 83, 189, 206
Programmable logic controller, 20
Prohibited food, 201
Protein denaturation, 25
Protonation, 150
Protozoa, 127, 134, 138, 190, 192
PS, *see* Polystyrene
PTM, *see* Pressure-transmitting medium

Pulse effect (P_E), 28, 31, 33, 39–40
Pulsed magnetic fields, 167–168, 186
Pulsemaster, 198
Pungent, 10, 156
PurePulse Technologies Inc., 11
PVDC, *see* Polyvinylidene chloride
PVOH, *see* Polyvinyl alcohol
Pyridoxine, 148
Pyrimidine, 66
Pyrolysis, 89

Q

Quadratic equation, 160
Quadratic polynomial model, 164
Quality parameters, 52–53, 62, 207
    quality assurance, 30
    quality attributes, 9, 29, 94, 133
    quality deteriorating enzyme, 89
Quantum theory, 143
Quartz, 68–69, 77–78, 90, 101, 109
Quasi-equilibrium, 105
Quaternary structure, 25, 37, 119
Quenched plasma, 111, 113, 121

R

Rad, 145
Radiant energy, 145
Radiant power, 72
Radiative recombination, 107, 116, 120–121
Radio frequency plasma, 109–110, 113, 115
    radioactive, 11–12, 145, 151–154
    radiochemical, 125, 133
    radiofrequency, 110, 114, 116, 121, 168, 201
    radioisotope, 151
    radiolysis, 145–151, 157, 193
    radiometer, 69
    radionuclide, 153
    radiosensitive, 149, 151, 157

Radius, 85, 96–97

Radura, 203

Ramp rate, 18, 23

Rancidity, 149, 162

Rarefaction, 85, 95, 101

Reactive species, 5, 7, 107–108, 111, 113, 115–116, 119, 131

Reactive nitrogen species, 108, 112

Reactor, 68–69, 77, 90, 96, 105, 111, 128–129, 139

Ready-to-eat, 8, 30, 208

Recombination, 105–8, 115–116, 120–121

Recontamination, 200

Reflection, 70, 78–79

Refraction, 70

Refrigeration, 195, 203

Renewable, 189, 195–196

Resistivity, 49, 59, 92, 96, 176–177, 180–181, 190, 193, 196

Resistors, 48, 175

Resonance, 90, 169–172

Respiration, 133, 174

Retailer, 12, 189, 206–207

Retort, 2

Reversible, 25, 46, 59, 190

Reynolds number, 201

RFP, *see* Radio frequency plasma

Ribosomes, 6

Ripening, 148–149

RNA, 5–7, 66

RNS, *see* Reactive nitrogen species

Rod-plate electrode system, 50

Rod-rod electrode system, 50–51

RTE, *see* Ready-to-eat

S

Sanitation, 13

Scale parameter, 31–33, 40, 50, 56–57, 116–117

Scanned beam electron accelerator, 152

Scattering, 70, 163

Schrödinger system, 185

Scratching, 120

Seafood, 8, 30, 157–159, 199, 203, 208

Secondary conformations, 172

Secondary metabolites, 114

Secondary protein, 47, 68

Secondary structure, 37, 102, 107, 116

Second-order kinetic models, 62

Seed germination, 114

Self-sustaining, 110

Semicontinuous, 18–19, 28–29

Semi-log, 59

Semipermeable, 46

Semisolid, 9, 43, 49, 54, 75–76, 191–192, 205

Sensory analysis, 156, 204

Sensory panel, 156

Shadowing effect, 72

Shape parameter, 22, 31–32, 40, 45, 50, 56–57, 68, 116–117, 143

Shear, 6, 86, 88, 96

Shelf-stable, 30, 206

Shock waves, 87–88, 96

Silent discharge, 109, 121

Silica, 68

Single-layered coil, 178

Singlet oxygen ($^1O_2$), 107

Sinusoidal waves, 176

Siren, 90

Smell, 10

SMF, *see* Static, static magnetic fields

Soaking, 108, 120

Social sustainability, 206–209

Sodium phosphate, 141

Softening, 149, 193

Solenoid, 174, 178, 180

Sonic frequency range, 85

Sonication, 87, 89, 98–99, 103–104, 191

Sonicator, 91, 95, 191

Sorbitol, 174

Spices, 10, 132, 149, 155–156, 164, 190

Spoilage enzyme inactivation, 9, 180

Spore-formers, 94, 151, 156

Sprout, 145, 149, 154–155, 177–179, 187

Stainless steel, 52, 55, 153

Stakeholder, 12, 189, 206–207

Standard, 53, 74, 172, 201–202, 209, 211

Starch, 149, 193

Static, 51, 167, 169, 186
  static magnetic fields, 167, 169, 171–172, 179
  static parallel plate, 51

Stationary, 25, 50, 113, 130, 177

Storage, 2, 62, 129–130, 153, 156

Stratosphere, 126

Streptomyces scabies, 182, 187

Subatomic particle, 169

Sub-lethally damaged, 88, 95–96, 102, 191

Substrate, 25, 147, 173

Sugar, 6, 26, 53, 94, 149, 206
  sucrose, 174

Sulfanyl group 'R-SH,' 68

Sulfhydryl, 5, 7, 26, 47, 128, 148

Superconducting, 174–175, 180–181, 186, 193

Supercooling, 174, 187

Superoxide, 107, 127, 134, 147

Surface disinfection, 10, 13, 76, 87, 115–116, 133, 192, 198, 205

Surface morphology, 25, 192

Suspension, 25, 44

Symbios Technologies, 198

Synergistic, 27, 67

T

TAC, *see* Total antioxidant capacity

Tail inactivation, 83

TAPB, *see* Total aerobic psychrotrophic bacteria

Taste, 53

TBC, *see* Total bacterial count

TC, *see* Total carotenoids

Tenderization, 10

Tenderness, 48, 87, 191–192

Teratogenic characteristics, 114

Tertiary structure, 25, 37, 68, 119, 147

Tesla, 11, 167, 169, 175, 177–179

Texture, 48, 108–109, 120, 133

TFC, *see* Total fungal count

Thermoacidophilic bacteria, 94, 104

Thermocouple, 22, 175

Thermodynamic equilibrium, 23, 105, 119–121

Thermodynamic nonequilibrium, 120

Thermoformed plastic, 200

Thermosensitive compounds, 4, 194

Thermosonication, 98–99

Thermostat, 175, 177

Thiamine, 148

Thiol groups, 68

Thorium, 14

Threonine, 147

Thymine, 66

Tissue, 144, 162, 169, 171, 174, 193

Titanium probe, 103

Titratable acidity, 94

TMP, *see* Transmembrane potential

Total aerobic psychrotrophic bacteria, 122

Total antioxidant capacity, 178–179; *see also* Antioxidant capacity

Total bacterial count, 178
Total carotenoids, 62
Total color change, 29, 74
Total fungal count, 178
Total polyphenol content, 53, 68,
    74–75, 104, 178–179
Total soluble solids, 26, 53,
    74, 89
Toxic, 54, 133, 190, 192
Toxins, 146, 156, 192–193, 195
TPC, *see* Total polyphenol content
Transducer, 89–90, 96–97
Transformer, 175
Translocation, 170–172, 180
Transmembrane potential, 44–46,
    49, 54, 59, 61
Transverse wave, 86, 100
Triglycerides, 149
Trimers, 148
Trioxygen, 10, 125
Tryptophan, 68, 147–148
TSS, *see* Total soluble solids
Tubers, 149
Tubular heat exchanger, 74
Tungsten, 68, 152
Turbidity, 70, 75, 94, 115
Turbine agitated reactor, 129, 139
Tyrosine, 68, 147–148

U

Ultra-HPP, *see* Hyperbaric pressure
    processing
Ultraviolet (UV-light), 7, 9, 65, 69,
    83–84, 141, 189
Uniform electric field, 43, 50, 55
Unipolar, 52
Universal gas constant, 27, 33, 36
Unpaired electrons, 107, 147, 173
    unpaired valence electrons, 107
Unpleasant odor-causing volatiles,
    128, 134, 192
Unprocessed food, 1

V

Vacuum, 65, 152, 172, 184
Valence electron, 106–107
Valorization, 195, 197, 211
van der Waals forces, 89, 102
van der Waals interactions, 7
Vaporization, 67, 112
Variation, 27, 49, 100, 146,
    154, 179
Vector, 167, 185
Vegetative, 113, 131, 177, 179
Velocity, 170–171, 173, 184, 186
Viable cells, 72
    viable cell counts, 182, 186
Virus, 127–128, 138, 151, 156–157,
    161, 192–193
Viscoelastic, 45–46, 54, 59, 61
Viscosity, 75, 162
Visible range, 66, 68, 73, 76,
    106–107
Vit-B$_1$, *see* Thiamine
Vit-B$_3$, *see* Niacin
Vit-B$_6$, *see* Pyridoxine
Vit-B$_{12}$, *see* Cobalamin
Volatile, 128, 134, 149, 192, 200,
    209

W

Water activity, 26, 88, 92, 151, 193
Water-binding capacity, 87, 191
Water-cooled magnetic coil, 175,
    181
Wave amplitude, 92, 96, 101
Wave frequency, 101, 143
Wave propagation, 100, 184
Wave pulses, 51
Wave spectrum, 144
Waveform, 50–52, 55, 57, 60, 167,
    180
Wb, *see* Weber
Weber, 167

Weibull distribution, 31, 40, 56, 113, 131
Weibull model, 56, 116
Welding arc, 105
Wheat germ, 164
Whistle reactor, 90, 96
White-box model, 185
Winding coils, 181
Wine, 52

X

Xe, *see* Xenon
Xenon, 66, 76–77, 79, 81, 191
X-ray, 11, 143–146, 151–153, 157, 160–161, 165, 185, 202

Y

Yeast and mold, 29, 37
Yield, 134
YM, *see* Yeast and mold
Yoke-driven mechanism, 21

Z

Zero-order, 135–136
Zeroth order, 62
Zone, 31, 38–39, 91, 113–114, 121
Zygosaccharomyces bacilii, 52–53